OXFORD MEDICAL PUBLICATIONS

Pharmacology of hydroxyethyl starch

Use in therapy and blood banking

Pharmacology of hydroxyethyl starch

Use in therapy and blood banking

JOHN MILTON MISHLER IV
A.B., Sc.M., D.Phil. (Oxon), M.R.C.Path.

Alexander von Humboldt Foundation Senior Research Fellow,
Laboratory of Tumour Immunology,
Medical University Clinic,
Cologne, West Germany

Presently

Chief, Blood Resources and Transplantation Branch,
Division of Blood Diseases and Resources,
National Heart, Lung, and Blood Institute,
National Institutes of Health,
Bethesda, Maryland, USA

OXFORD
OXFORD UNIVERSITY PRESS
NEW YORK TORONTO
1982

Oxford University Press, Walton Street, Oxford OX2 6DP
London Glasgow New York Toronto
Delhi Bombay Calcutta Madras Karachi
Kuala Lumpur Singapore Hong Kong Tokyo
Nairobi Dar es Salaam Cape Town
Melbourne Auckland
and associate companies in
Beirut Berlin Ibadan Mexico City

British Library Cataloguing in Publication Data
Mishler, John Milton
Pharmacology of hydroxyethyl starch.
– (Oxford medical publications)
1. Chemotherapy 2. Hydroxyethyl starch
I. Title
615'.3'3 RM666.H/
ISBN 0-19-261239-5

Library of Congress Cataloging in Publication Data
Mishler, John Milton, 1946–
Pharmacology of hydroxyethyl starch.
(Oxford medical publications)
Bibliography: p.
Includes index.
1. Hydroxyethyl starch – Physiological effect.
2. Hydroxyethyl starch – Therapeutic use. 3. Blood plasma substitutes. I. Title. [DNLM: 1. Starch –Pharmacodynamics. 2. Starch – Therapeutic use. 3. Plasma volume. 4. Blood banks. QU 83 M678p]
RM666.H87M57 615'.718 81-18923
ISBN 0-19-261239-5 AACR2

Set by Hope Services, Abingdon
Printed in Great Britain
at the University Press, Oxford
by Eric Buckley,
Printer to the University

Foreword

by

Colin R. Ricketts

MRC Industrial Injuries and Burns Unit, Birmingham Accident Hospital, Birmingham, England

Those with experience of other colloids for clinical use may very well ask 'why bother with hydroxyethyl starch?' Compared with other colloids, hydroxyethyl starch offers a wide spread of molecular weight with more nearly spherical molecules. There is thus a lower viscosity for any given molecular weight. Molecular size and shape influence the distribution of colloid molecules in the body. Hydroxyethyl starch is slowly hydrolysed by α-amylase present in plasma, the rate of hydrolysis depending upon the number and pattern of hydroxyethyl group substitution in the molecule. In addition, there is the very interesting but not wholly understood physical property of protecting red cells against the damaging effects of freezing and thawing. Clearly, hydroxyethyl starch has fascinating possibilities which go beyond those of other colloids yet investigated.

Dextran and gelatin have been widely used as plasma volume expanders and this has led to much scientific effort to understand fully the nature of these materials and their biological behaviour. This is particularly so in the case of dextran which must be one of the most widely researched colloids. Hydroxyethyl starch has been used as a plasma volume expander but perhaps its most exciting uses are for the protection of red cells against freezing and for the separation of leucocytes from blood.

This book brings together for the first time all that is known about hydroxyethyl starch. The first chapter on the molecular structure and chemistry brings out in detail the features referred to above, while appendices give full experimental information on the analytical methods used. The chapter on catabolism, excretion and tissue storage of hydroxyethyl starch is a comprehensive review of the large literature on this aspect. Japanese work on hydroxyethyl starch has been more extensive than is generally realized and due attention has been given to it. Other chapters deal with all aspects of the effects of hydroxyethyl starch infusions in man and animals. It is noteworthy that in man the plasma α-amylase concentration rises after the infusion of hydroxyethyl starch and it seems possible that this is due to the adsorption of α-amylase onto the hydroxyethyl starch molecule with the complex remaining in the bloodstream, instead of the α-amylase being excreted in the urine. Use of hydroxyethyl starch as a plasma volume expander is fully covered. Details are given of the separation of leucocytes from blood and there is much information on the use of hydroxyethyl starch for storing whole human blood in the frozen state.

Dr Mishler has personally done research on the various applications of hydroxyethyl starch in several countries including the USA, Britain, and Germany. His comprehensive account will be read with great interest by all concerned with the many applications of hydroxyethyl starch.

Foreword

by

David Mason Robinson
Office of the Director,
National Heart, Lung, and Blood Institute, National Institutes of Health,
Bethesda, Maryland, USA

Plasma expander, cryoprotective compound, diluent for heart–lung machines, agent in cytapheresis and in the production of leucocyte-free red blood cells; so the list of uses of the hydroxyethyl starches has grown steadily since first introduced by Wiedersheim in 1957. Probably the full potential of these safe and effective volaemic colloids has not yet been fully realized, but by now a very substantial literature of hydroxyethyl starch has accumulated and the collection of this into a comprehensive work of reference is both timely and desirable. We should be grateful to John Mishler, himself the author of much modern experimental work on hydroxyethyl starch, who has now provided such a welcome monograph, prepared in a thorough and scholarly fashion.

The sequence and organization of the chapters of this book are logical and the detailed subdivisions most helpful. The reader may choose to study only those sections relevant to his own interest and these can be identified with ease and facility. Each section can be read as an isolated entity, without recourse to the rest of the text (except where expressly directed to some other section for more detail or for clarification) making this a convenient and valuable source of information. The literature has been surveyed comprehensively, providing a lucid synthesis from a complete and up-to-date bibliography. As a point of departure for all types of medical or biological applications, ranging from clinical practice to studies at the molecular level, this work is truly outstanding.

The safety and efficacy of hydroxyethyl starch in animals and man no doubt derives in part from the compact, highly branched structure of the molecules, so reminiscent of natural glycogen. Such ready tolerance is emphasized by work reported here, showing hydroxyethyl starch neither to elicit antibody formation, nor to cause significant release of histamine. Under these circumstances, increasing application of hydroxyethyl starch to blood-related problems seems highly likely and workers in this area, faced with the design of experimental procedures or clinical protocols, will now be able to operate from the vastly improved standpoint that is provided by John Mishler's complete and critical review. As a consequence of the availability of this monograph, it is clear that the pace of scientific enquiry will inevitably quicken.

To my son Joshua Evan Mishler, and my wife Sigrid Ruth Elisabeth Fischer-Mishler.

Success is never final and failure never fatal. It's courage that counts

Quoted in *Success Unlimited*

Preface

Poetically speaking '*a journey of one-thousand miles begins with a single step*', and so it may be said realistically that the first step in the development of the volaemic colloid hydroxyethyl starch began with the early discoveries of Ziese (1934, 1935). His work elucidated the inhibitory effect of hydroxyethylation on α-amylase mediated catabolism of starch. This initial or first step in the development of hydroxyethyl starch lay dormant for over 20 years, however, until Wiedersheim (1957) saw the practicality of using a less rapidly degraded form of starch as a means to restore a diminished plasma volume in cats subjected to haemorrhagic shock. The studies of Wiedershiem thus married together for the first time the theory of hydroxyethylation of starch and its logical extension, its use as a volaemic colloid.

When starch is hydroxyethylated, two variables – namely molecular weight (MW) of the parent starch molecule and the degree of hydroxyethyl group substitution (MS) – can be exploited in developing volaemic colloids that survive in blood for varying lengths of time. It was these two variables of hydroxyethyl starch and their application, first in animal models and later in man, that were investigated next by Walton and his colleagues in the early 1960s. This group of investigators 'fine-tuned' hydroxyethyl starch, testing various combinations of MW and MS to establish the most desirable characteristics required in a volaemic colloid. They also studied the effect of this material on organ function, coagulation, antigenicity, and rheology. From these investigations conducted by Walton and his colleagues, the first species of hydroxyethyl starch (HES 450/0.70) was developed for testing in man.

Since these early studies, numerous investigators have reviewed the efficaciousness of several species of hydroxyethyl starch in volume resuscitation (Fiala 1979; Gryszkiewicz 1978; Köhler 1978; Mishler 1980*a*, *d*; Polushima 1980; Thompson 1974, 1978). Although these reviews are useful, they do not systematically detail the many facets of the pharmacology of the various species of hydroxyethyl starch. I have, therefore, attempted in this monograph to complete the next '*950 miles of the journey*' in our understanding of the many features of this most fascinating material. I have attempted to update our knowledge on the *in vitro* and *in vivo* catabolism and excretion of hydroxyethyl starch with special emphasis on its storage in various tissues. In the subsequent chapters, I have isolated the effect of dose of material injected with corresponding host response. In this manner, it should be possible to predict the risk to the host as it relates to the dose of hydroxyethyl starch injected.

Hydroxyethyl starch was originally developed to increase and sustain a deficient plasma volume, but over the past decade this material has found use in several areas of interest to blood bankers. These new applications are also dealt with in this monograph.

The formidable task of synthesizing our present knowledge of hydroxyethyl starch would have been impossible without the assistance of colleagues. Trevor Greenwood and Donald Muir have, with their extensive knowledge of the physicochemical properties of hydroxyethyl starch, prepared Chapter 1. Edward D. Allen, Joseph C. Fratantoni, David M. Robinson and Brenda J. Slade skilfully reviewed portions of the monograph, and I thank them for a job well done. I should also like to extend my warm thanks to colleagues who furnished reprints of their work on hydroxyethyl starch: E.D. Allen, H. Bergmann, D. French, M. Fujimori, Y. Goto, J.P. Hester, B. Hölscher, H. Köhler, F.J. Lionetti, W. Lorenz, H.G. Merkus, K. Messmer, D.E. Pegg, W. Richter, P. Safar, M.W. Scheiwe, J.A.R. Smith, W. Sonntag, and C. Watzek.

My secretary Ms Kathy White completed the task of putting this monograph together, and without her skill and good humour, I would have faced an impossible task. The editorial staff of the Oxford University Press have given much-appreciated support in bringing this work to press.

Bethesda, Maryland JMM IV
25 September 1981

Contents

1. The structure and chemistry of hydroxyethyl starch

1.1. INTRODUCTION

Methods are now available to produce a very wide range of material which falls within the generic definition of hydroxyethyl starch. The most important properties of hydroxyethyl starch, the level of substitution and the viscosity, are controlled by methods which are readily applicable. Also, within certain limits, the position of hydroxyethylation on the polymer backbone may be manipulated at will. However, it is equally clear that polysubstitution reactions occur with whatever reaction scheme has been used to date, and that the level of polysubstitution increases disproportionately at higher overall levels of substitution (Merkus *et al.* 1977).

Hydroxyethyl starch has been widely characterized by several techniques. It is now feasible not only to measure *molar substitution*, but also *amount* and *pattern* of *substitution* by accurate and elegant analytical techniques. More recent advances have made possible routine and detailed analysis of the chemical structure.

Similar advances in analytical technique allow ready evaluation of not only the molecular weight and viscosity of samples of hydroxyethyl starch but also the molecular weight distribution. Perhaps the most important physical attribute of hydroxyethyl starch is that its branched structure results in a relatively low viscosity for a given molecular weight. Thus, the hydroxyethyl starch molecule has advantages over dextran in many biological applications.

The control of the enzymic degradation of hydroxyethyl starch may be achieved by two independent mechanisms. That is, by control of degree of substitution and also, but to a lesser extent, by manipulation of the substitution pattern.

The technology and quality control methods are now available to produce a very wide range of molecular species. The choice must now be made by the clinician or scientist either to adopt a compromise or to opt for different specialist types of hydroxyethyl starch in much the same way that various types of dextran are available. This chapter will deal with the physicocochemical preparation of hydroxyethyl starch. In subsequent chapters, the various clinically-tested species of hydroxyethyl starch will be described in detail in studies in both man and animals.

1.2. PREPARATION OF HYDROXYETHYL STARCH

Hydroxyethyl starch is a relatively simple polymer to prepare; it is formed by the reaction between ethylene oxide and amylopectin in the presence of an alkaline catalyst (Ziese 1934, 1935). In aqueous solution the reaction probably occurs by either one of the nucleophilic substitution sequences shown below:

$$\underset{\text{Starch}}{\text{R-O-H}} \xrightarrow{\text{base}} \text{R-O}^- + \text{H}^+ + \underset{\text{Ethylene oxide}}{\overset{\text{O}}{\text{CH}_2\text{-CH}_2}} \xrightarrow{\text{base}} \underset{\text{Hydroxyethyl starch}}{\text{R-O-CH}_2\text{CH}_2\text{OH}} \tag{1.1}$$

$$\underset{\text{Starch}}{\text{R-O-H}} \xrightarrow{\text{base}} \text{R-O}^{\delta-} \text{---} \text{H}^{\delta+} + \underset{\text{Ethylene oxide}}{\overset{\text{O}}{\text{CH}_2\text{-CH}_2}} \xrightarrow{\text{base}} \underset{\text{Hydroxyethyl starch}}{\text{R-O-CH}_2\text{CH}_2\text{OH}} \tag{1.2}$$

Irrespective of the two reaction schemes (1.1) and (1.2), partial or total ionization of the hydroxyl groups on the polymer backbone is a necessary prerequisite for the reaction to occur. Side-reactions may also occur, the most important one being the hydrolysis of ethylene oxide to ethylene glycol:

$$\underset{\text{Ethylene oxide}}{\overset{\text{O}}{\text{CH}_2\text{-CH}_2}} + \text{H}_2\text{O} \xrightarrow{\text{base}} \underset{\text{Ethylene glycol}}{\text{HOCH}_2\text{-CH}_2\text{OH}} \tag{1.3}$$

Ethylene glycol is toxic and must be removed from hydroxyethyl starch by repeated solvent extraction (Schoch 1965). Despite such precautions, small quantities of ethylene glycol have been detected in hydroxyethyl starch intended for clinical use (de Belder *et al.* 1976).

Hydroxyethyl starch is an extremely complex polymer because: (i) the parent amylopectin molecule is heterogeneous both in chemical and physical properties; and (ii) hydroxyethylation can proceed at several different sites on the amylopectin molecule, influencing blood-persistence characteristics (see Section 2.2.2).

1.2.1. STARCH MATERIAL FOR THE PREPARATION OF HYDROXYETHYL STARCH

The starch most commonly used for the preparation of hydroxyethyl starch is a waxy species of either maize or sorghum. Starches exist as semicrystalline, individual granules of polymeric material. In waxy starches, the predominant glucan polymer is *amylopectin* (*c.* 98 per cent), but the amount may vary between species, and even between cultivars of the same species (Banks *et al.* 1973).

The predominant linkage between the glucose residues of the amylopectin chains is the α-1:4-bond, but chains of between 16 and 25 residues (depending on the source of the amylopectin) are attached by α-1:6-bonds. The resultant highly branched structure is thought to be essentially random (Meyer *et al.* 1941) and of the form shown schematically in Fig. 1.1. Doubts have been cast on the adequacy of this model (Gunja-Smith *et al.* 1970), but conclusive evidence is not yet available. Notwithstanding these doubts concerning the chemical structure of amylopectin, the physical form assumed by the polymer in solution is also unusually complex. There is, for example, strong evidence that the polysaccharide structure is much more extended than would be expected from studies of its more highly branched analogue glycogen. For example, the viscosity of amylopectin is about 15 times that of glycogen of comparable molecular weight (see Banks and Greenwood (1975) for an extensive review of the hydrodynamic behaviour of amylopectin).

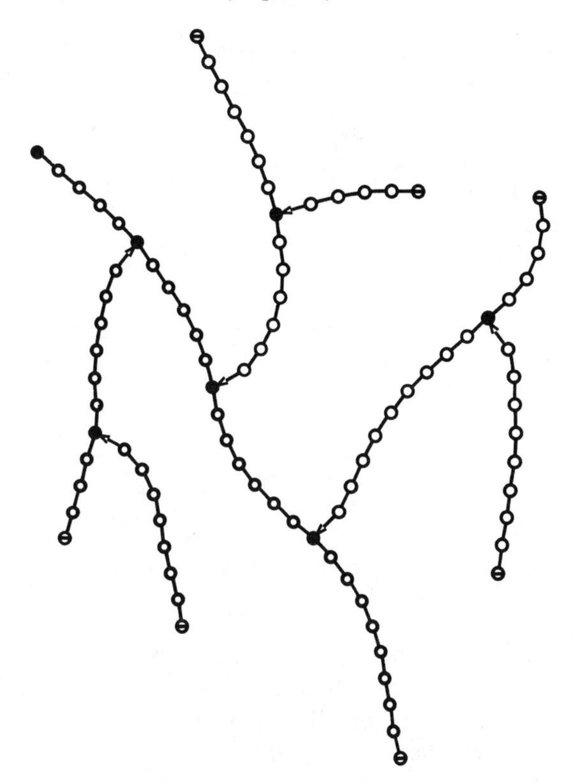

Fig. 1.1. A schematic representation of the randomly branched structure of amylopectin. Non-reducing chain end (⦶); reducing chain-end (●); α-1:4 bond (○ – ○); and α-1:6-bond (○ → ●).

1.2.2. THE HYDROXYETHYLATION REACTION

This occurs at the three sites on each (unbranched) glucose residue in the amylopectin chain that are *initially* available for substitution, i.e. the hydroxyl groups at carbon atoms 2, 3, and 6 (C2, C3, C6) of the glucose ring (Fig. 1.2).

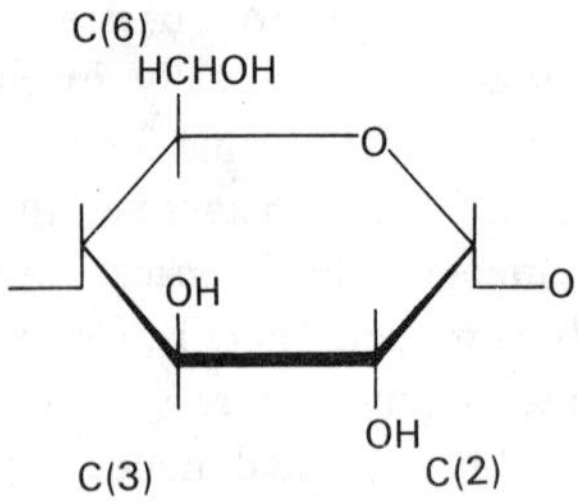

Fig. 1.2. Anhydroglucose residue in hydroxyethyl starch. The numbering convention is clockwise as shown.

In addition, each newly introduced hydroxyethyl group also carries a hydroxyl group, which may itself react further:

$$\underset{\text{Hydroxyethyl starch}}{R\text{-}OCH_2CH_2OH} + \overset{O}{CH_2\text{-}CH_2} \xrightarrow{\text{base}} R\text{-}OCH_2CH_2OCH_2CH_2OH \qquad (1.4)$$

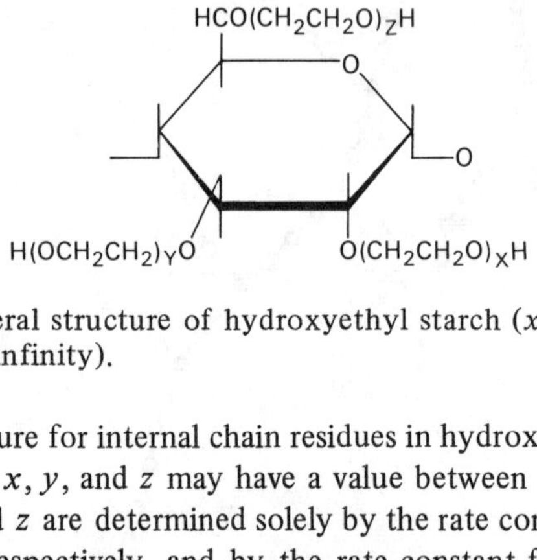

Fig. 1.3. The general structure of hydroxyethyl starch (x, y, and z have values between zero and infinity).

The general structure for internal chain residues in hydroxyethyl starch is shown in Fig. 1.3, where x, y, and z may have a value between zero and infinity. The values of x, y, and z are determined solely by the rate constants for reactions at C2, C3, and C6 respectively, and by the rate constant for the polymerization reaction (eqn. 1.4).

1.2.3. MODEL FOR THE REACTION

It is convenient to use the following mathematical model (Banks *et al.* 1972) to understand the manner in which the substitution pattern of hydroxyethyl starch is controlled. Each glucose residue has a hydroxyl group available for reaction at C2, C3, and C6. If it is assumed that the reaction at any given site is independent of the status of other reactive sites, and that the rate constant is proportional to the number of unreacted sites, e.g. a first-order reaction, we may write (Spurlin 1939):

$$\frac{\mathrm{d}S}{\mathrm{d}t} = k(1-S) \tag{1.5}$$

where S is the fraction of available sites which have reacted and k is the rate constant for the substitution reaction. Integration of eqn (1.5), with the constraint that $S = 0$ when $t = 0$, gives:

$$\ln(1-S) = kt, \tag{1.6}$$

or, expressed in exponential form

$$S = 1 - \mathrm{e}^{-kt}. \tag{1.7}$$

Each of the available hydroxyl groups on starch (or its derivatives) will react in the manner shown by eqn 1.7, but there is no reason to suppose that the rate constants for all three sites will be identical. We may recognize this fact by writing three equations:

$$S_2 = 1 - \mathrm{e}^{-k_2 t} \tag{1.8}$$

$$S_3 = 1 - \mathrm{e}^{-k_3 t} \tag{1.9}$$

$$S_6 = 1 - \mathrm{e}^{-k_6 t} \tag{1.10}$$

where S_2, S_3, and S_6 are the extent of substitution at C2, C3, and C6, respectively, and k_2, k_3, and k_6 are the appropriate rate constants.

The rate of the polymerization reaction (eqn 1.4) is proportional to the fraction of hydroxyl groups already substituted. Therefore,

$$\mathrm{d}S_\mathrm{p} = k_\mathrm{p}\,(S_2 + S_3 + S_6)\,\mathrm{d}t. \tag{1.11}$$

S_2, S_3, and S_6 may be replaced from eqns 1.8, 1.9, and 1.10 to give

$$\mathrm{d}S_\mathrm{p} = k_\mathrm{p}\,[(1-\mathrm{e}^{-k_2 t}) + (1-\mathrm{e}^{-k_3 t}) + (1-\mathrm{e}^{-k_6 t})]\,\mathrm{d}t. \tag{1.12}$$

If, for the moment, S_3 and S_6 are ignored, then

$$\mathrm{d}S_\mathrm{p} = k_\mathrm{p}\,S_2\,\mathrm{d}t = k_\mathrm{p}\,(1-\mathrm{e}^{-k_2 t})\,\mathrm{d}t. \tag{1.13}$$

Integration then yields the relation

$$S_\mathrm{p} = k_\mathrm{p}\,(t - \frac{1}{k_2}\,\mathrm{e}^{-k_2 t} + C) \tag{1.14}$$

where C is the integration constant. Applying the boundary condition that $S_p = 0$ when $t = 0$, it follows that

$$C = -\frac{1}{k_2}.$$

Substitution for C in eqn 1.14 yields

$$S_p = k_p t - \left(\frac{1}{k_2}\right) \cdot (1 - e^{-k_2 t}). \qquad (1.15)$$

Returning to eqn 1.10, the terms containing k_3 and k_6 may be integrated in an analogous manner to yield:

$$S_p = k_p \ \{3t - [\left(\frac{1}{k_2}\right) \cdot (1 - e^{-k_2 t}) + \left(\frac{1}{k_3}\right) \cdot (1 - e^{-k_3 t}) + \left(\frac{1}{k_6}\right) \cdot (1 - e^{-k_6 t})]\}. \qquad (1.16)$$

There is no simple way in which the various rate constants can be determined, but by comparison of the amounts of substitution at various sites and by application of the calculations shown above it is possible to establish the ratios between the various rate constants. An example of such a calculation will be presented in Section 1.3.3. below.

1.2.4. METHODS USED FOR HYDROXYETHYLATION

A wide variety of catalysts have been used commerically for the preparation of hydroxyethyl starch, including inorganic salts, inorganic alkalis, and certain organic bases (see Broderick 1960; Kerr and Faucette 1956; Kesler and Hjermstead 1950*a*,*b*,*c*; Sadota 1974, 1975). A number of reaction systems have been designed to produce essentially granular products by retention of at least some of the characteristics of the native starch. This may be achieved by inhibition of swelling of the starch: (i) by limiting the amount of water present in the reaction system (Caldwell and Martin 1957; Greenwood *et al.* 1977; Kerr and Faucette 1956); (ii) by the addition of inorganic salts; or (iii) by carrying out the reaction in alcoholic media (Broderick 1954; Hjermstead 1959; Kesler and Hjermstead 1958). Hydroxyethyl starch has also been prepared by the reaction of ethylene chlorohydrin with pregelatinized starch in pyridine (Mima and Gokoyama 1970) and with starch alkoxide in pyridine (Gaver 1950).

Although it is by no means certain, there is evidence that most of the hydroxyethyl starch which has been evaluated in clinical trials was prepared by the method described by Schoch (1965). The technique is shown diagrammatically in Fig. 1.4 and comprises the following nine operations:

1. The starch receives a preliminary alkali treatment to reduce the level of

protein and natural pigment which contaminate the granular starch.

2. Acid hydrolysis with dilute acid is then used to reduce the molecular weight of the final product.

3. The hydrolysed starch (still in the granular form) is then dissolved in aqueous sodium hydroxide.

4. The resulting solution is then treated with ethylene oxide in a pressurized reaction vessel.

5. After completion of the hydroxyethylation step the solution is decolorized by carbon treatment.

6. Haze and suspended particulate material is removed by filtration.

7. The resulting clear solution is spray-dried to yield a white powder.

8. Glycols formed as by-products of the hyroxyethylation (see eqn 1.3) are then removed by extraction of the powder with aqueous acetone.

9. Finally, the acetone is removed by drying.

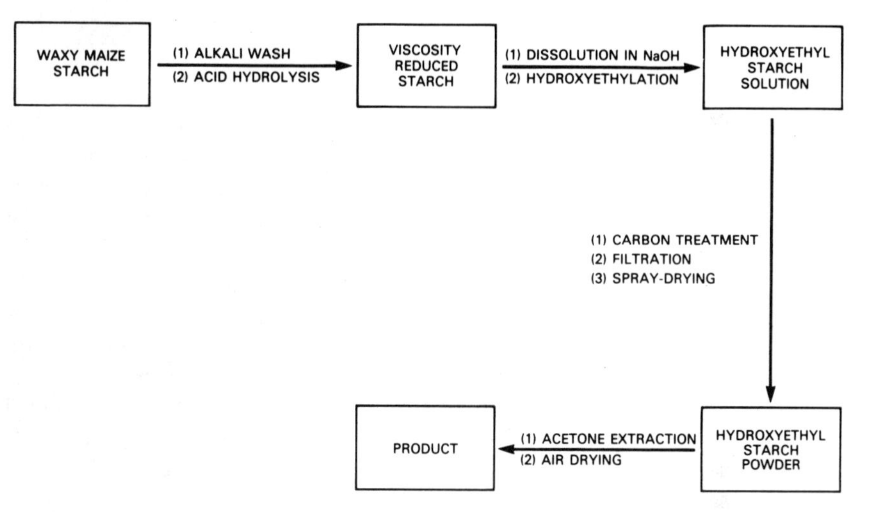

Fig. 1.4. Preparation of hydroxyethyl starch according to Schoch (1965).

The process is lengthy and requires considerable amounts of sophisticated equipment. In addition, hydroxyethyl starch powders produced by this method suffer from the disadvantage that the material is contaminated with sodium chloride (1.2–2.8 per cent) and cannot be used if reversible agglomeration of erythrocytes is required for post-thaw processing of human erythrocytes (Greenwood *et al.* 1974).

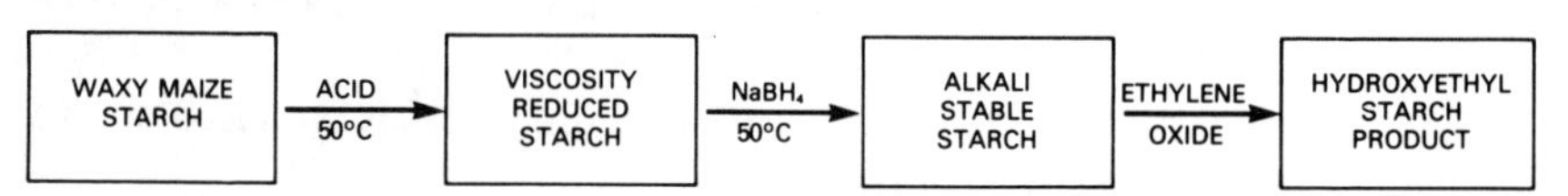

Fig. 1.5. A schematic representation of the process for the manufacture of hydroxyethyl starch.

An alternative procedure has been developed which overcomes many of the problems associated with the Schoch procedure (Greenwood *et al.* 1975, 1978).

The method comprises three steps and is shown schematically in Fig. 1.5. The original granular structure and form of the starch are retained throughout the reaction scheme by using a series of heterogeneous reactions. The steps are summarized below:

STEP 1 – ACID MODIFICATION

Waxy maize is treated with dilute hydrochloric acid (2 mol/l) at a temperature below the gelatinization temperature of the starch. After the desired reaction time (2-4 h at 50 °C) the starch–acid slurry is rapidly cooled, then neutralized by the addition of sodium bicarbonate.

After repeated differential centrifugation to remove small quantities of dark-coloured material formed in side-reactions, the acid-modified starch is slurried in distilled water, washed, and air-dried. A white free-flowing granular powder is produced.

STEP 2 – STABILIZATION

During acid modification the hydrolysed glucosidic bonds yield new, labile reducing end-groups which decompose either at high temperatures (such as during sterilization procedures) or in the presence of alkali to give yellow or brown products. The labile end-groups can be stabilized by conversion to the sorbitol analogue. It is thought that the small amount (~ 3 per cent) of these groups would be biologically acceptable because the free sugar sorbitol is already used in some intravenous solutions (Irving and Rushman 1971).

Acid-modified starch is stirred in warm water (48–50 °C) and sodium borohydride added (1 per cent w/v). After 5 minutes the reaction is terminated by adding glacial acetic acid. The white granular starch is then washed with distilled water and air-dried. The product has under 10 per cent of the original reducing powder and is stable when treated with hot, dilute alkali.

STEP 3 – DERIVATIZATION

Hydroxyethyl groups are introduced into the stabilized starch by treatment with ethylene oxide in the presence of a basic catalyst. Dry, acid-modified starch is stirred in distilled water containing iso-propanol and sodium hydroxide and, after cooling the slurry to 2 °C, ethylene oxide is added (the amount of ethylene oxide added governs the final level of substitution). After reaction at 37 °C for 24 h in a sealed vessel, the reaction is stopped by adding glacial acetic acid. The product is washed with iso-propanol, then with aqueous acetone and aqueous ethanol to remove residual glycols and sodium acetate.

The properties of a range of hydroxyethyl starch products produced by these techniques are shown in Table 1.1.

As pointed out in eqns 1.1 and 1.2, hydroxyethylation proceeds via a nucleophilic substitution reaction in which the hydroxyl group on the polymer backbone

TABLE 1.1 *Properties of hydroxyethyl starch*

Property	Sample Code W18*	W19*	W20
Moisture (%)†	0.50	2.80	6.50
Lipid (%)†	0.10	0.10	0.06
Protein (%)†	0.06	0.05	0.04
Ash (%)†	0.13	0.12	0.16
Molar substitution†	0.70	0.71	0.71
Viscosity (cp)‡	200	110	120
Colour §	Pale Yellow	Pale Brown	White

* No stabilization step was carried out in these samples
† Measured as in Greenwood *et al.* (1975)
‡ Viscosity of 35 per cent w/v solution
§ Colour of dry product

is partially or totally ionised. It is known that the strength of the base can have a considerable effect on the *extent* of hydroxyethylation of starch during homogeneous reaction of ethylene oxide with either amylose or amylopectin. For example, when solutions of waxy maize amylopectin are prepared in different strengths of potassium hydroxide and various quantities of ethylene oxide added, the property of the basic reaction medium has a profound effect (Table 1.2). This effect is particularly marked when hydroxyethylation is carried out at pH 12.0 in the presence of added salt. In this environment the structure of starch components assumes a different conformation than that normally found in solution and some evidence suggests that the starch has some helical characteristics (Banks and Greenwood 1975). Ionized groups may be necessary to stabilize such helical structures at pH 12.0 in the presence of salt, and such ionic groups on the starch would almost certainly be highly reactive towards ethylene oxide (eqns 1.1 and 1.2). Although it is also probable that concomitant changes in the site of substitution also occur in this unusual reaction medium, this aspect of the reaction has not received detailed investigation.

TABLE 1.2 *The derivatization of amylopectin in alkaline solutions – the effect of the basic medium composition on the efficiency of hydroxyethylation*

Added ethylene oxide (ml)*	Molar substitutions in media containing:† KOH (0.01 mol/l)	KOH (0.01 mol/l) + KCl (0.3 mol/l)	KOH (1.0 mol/l)
6	0.35	0.60	0.49
12	0.56	1.08	0.84
18	0.75	1.35	1.16
24	0.84	–	1.41
30	0.96	–	–

* Amount of ethylene oxide added to an amylopectin solution (2 per cent w/v) containing base and/or salt as shown
† The molar substitution was measured at completion of the reaction with amylopectin

1.3. CHARACTERIZATION OF THE CHEMICAL STRUCTURE OF HYDROXYETHYL STARCH

1.3.1. MOLAR SUBSTITUTION

The single most important characteristic of a hydroxyethyl starch is the extent of substitution, or *molar substitution* (MS), defined as the average number of hydroxyethyl groups reacted per anhydroglucose residue. This may be calculated as

$$MS = \frac{W_H}{1 - W_H} \times \frac{162}{44} \qquad (1.17)$$

where W_H is the weight fraction of hydroxyethyl group in the polymer.

Although a number of techniques have been reported for measurement of MS there is only a single primary method of analysis. The hydroxyethyl ether group is hydrolysed with hydroiodic acid under a variety of conditions and the reaction products derived from this group are subsequently estimated.

With constant boiling hydroiodic acid at atmospheric pressure, cleavage of the hydroxyethyl ether linkage yields a mixture of ethyl iodide and ethylene. The method of Morgan (1946) and the later modification of Lortz (1956) with this reaction and the hydrolysis products were subsequently estimated by volumetric techniques. The method works well in practice but MS can at best be estimated to ± 5 per cent. The replicability of the technique can be improved (to ± 1 per cent) by using a modified ethylene trap as described by Boxall *et al.* (1974).

Alternative systems may be used in which a single hydrolysis product – ethyl iodide – is formed. Van der Bij (1967) used 70 per cent hydroiodic acid and carried the hydrolysis out under pressure. The ethyl iodide so liberated was then estimated by gas–liquid chromatography after solvent extraction. Hodges and his colleagues (1979) employed a similar method and used adipic acid to catalyse the hydroiodic acid cleavage of the hydroxyethyl ether linkage.

All the above methods require calibration against a primary standard and de Belder *et al.* (1972) carried out an extensive examination of suitable materials. These investigators concluded that 1,2-0-(ethylene)- α D-glucopyranose was a highly satisfactory standard.

Another approach was adopted by Anderson and Zaidi (1963). After hydrolysis of hydroxyethyl starch under conditions similar to those used by Morgan (1946), the gaseous reaction products (ethyl iodide and ethylene) were estimated by quantitative infra-red spectroscopy.

Tai *et al.* (1966) used a secondary method of analysis in which the acetaldehyde produced from the controlled hydrolysis of hydroxyethyl starch was estimated by gas–liquid chromatography. However, although rapid, the technique has limitations as it was calibrated by the use of samples whose molar substitution had been measured by the technique of Lortz (1956).

It is not usually appreciated that under the best possible conditions the accuracy of the determination of the hydroxyethyl ether content of hydroxyethyl starch is limited by side-reactions during hydrolysis. These side-reactions result from oxidation of the polymer backbone under the extreme conditions used to obtain complete cleavage of the ether linkages.

1.3.2. DEGREE OF SUBSTITUTION

Although of primary importance, measurement of MS alone provides no information about the distribution of hydroxyethyl substituents between the glucose residues of the amylopectin chains. The classical structure for hydroxyethyl starch in medical reports (Cerny *et al.* 1965, 1967; Russell *et al.* 1966; Thompson *et al.* 1962) was oversimplified, for it was thought that the reaction occurred almost entirely by monosubstitution at C6. Polymeric side-chains or residues with multiple substitution were not considered. This oversight appears to be a result of early reports of Schoch (1963), even though later it was thought that the derivatization process was random in nature (Schoch 1965). As pointed out in Section 1.2.3, substitution is probably complex (this assumption is validated later in this section) and polymeric side-chain formation is not only possible but probable if k_p is finite. Hence, measurement of MS will overestimate the number of anhydroglucose residues on the polymer chains which are substituted. In recognition of this problem a second parameter – *degree of substition* (DS) – is defined. The DS may be calculated from:

$$\mathrm{DS} = 1 - \frac{0.9\ W_\mathrm{G}\ (162 + 44\ \mathrm{MS})}{162\ W_\mathrm{p}} \tag{1.18}$$

where W_G is the weight of *free* glucose in a polymer sample of weight W_p.

Banks *et al.* (1973) described a simple technique for measurement of DS, as follows:

ASSAY METHOD

Hydroxyethyl starch is dried *in vacuo* at 70 °C overnight then dissolved in water to give a known concentration in the range 1-3 mg/ml. An equal volume of sulphuric acid (1.5 mol/l) was added, and the polysaccharide hydrolysed for 3 hours on a boiling water-bath. This procedure has been shown to achieve complete hydrolysis of the glycosidic bonds in amylose and amylopectin without any concomitant acid reversion of the liberated sugar. The ether links of the derivatized glucose residues are resistant to this relatively mild hydrolysis. The samples are cooled to 22-25 °C, neutralized by the addition of a predetermined amount of potassium hydroxide (1.0 mol/l), and diluted in a graduated flask with water. Dilution is such that a free glucose content in the range 5-40 μg/ml is achieved. An aliquot of the neutral solution (1.0 ml) is taken for analysis by the coupled glucose oxidase-peroxidase-chromogen assay system (Banks *et al.*

1970*a,b*). Control experiments with the various monosubstituted hydroxyethyl derivatives show that only free glucose could react with the enzyme–chromogen system.

The DS, however, can be measured by other, more complex methods. For example, Yoshida *et al.* (1973) used an alternative method where after hydrolysis, the hydrolysate was de-ionized with ion-exchange resin, the sugar mixture derivatized, and the derivatives separated and quantified by analytical gas-liquid chromatography. By this method, substituted glucose in the reaction mixture was identified and its concentration estimated.

1.3.3. DEFINITION OF THE SUBSTITUTION PATTERN OF HYDROXYETHYL STARCH

Having described methods for the measurement of the total amount of hydroxyethyl groups reacted with starch (MS), and the distribution of unsubstituted anhydroglucose residues (1–DS), the final requirement for complete chemical characterization of hydroxyethyl starch is that the distribution of substituted groups on the glucose residue be defined.

A number of methods have been used to define substitution pattern. The first technique uses degradation of the polymer itself. Of the available techniques, periodate oxidation (Banks *et al.* 1971; Husemann and Kafka 1960; Srivastava and Ramalingam 1967) has been widely used. The periodate ion oxidizes hydroxyl groups on contiguous carbon atoms, which in the case of starch means that oxidation occurs at positions C2 and C3. Therefore, if periodate uptake by hydroxyethyl starch is significantly lower than for starch, there is strong evidence that hydroxyethylation has occured at either C2 or C3 or at both sides. In fact, Banks *et al.* (1971), Husemann and Kafka (1960), and Srivastava and Ramalingam (1967) all observed a reduction in periodate uptake with samples of hydroxyethyl amylose and hydroxyethyl starch. Thus in these samples, it may be concluded that substitution did not occur primarily at C6.

Husemann and Kafka (1960) supported this finding by use of another direct technique. They tritylated hydroxyethyl starch (i.e. formed the trityl ether) and found that as MS increased, the extent of tritylation increased (tritylation is a reaction which is predominantly associated with primary hydroxyl groups (e.g. $-CH_2OH$ groups) which occur only at C6 or as $R–OCH_2CH_2OH$ on newly introduced hydroxyethyl groups). Thus there was confirmatory evidence that hydroxyethylation occurred principally at the secondary hydroxyl groups (C2 and C3) in starch.

A more widely-used technique for establishing the substitution pattern uses hydrolysis of hydroxyethyl starch to a mixture of its constituent monosaccharides, followed by estimation of the hydroxyethyl glucose derivatives either by paper chromatography (Bollenback *et al.* 1969) or by gas–liquid chromatography after suitable sample preparation (Banks *et al.* 1971; de Belder and Norrman 1969; Lott and Brobst 1966; Merkus *et al.* 1977; Norrman 1969; Yoshida *et al.* 1973).

A further extension of this concept was used by Mourits *et al.* (1976) in which gas–liquid chromatographic analysis was coupled with analysis of the fractions by mass spectrometry. This technique thus enables the identification of not only monosubstituted glucose residues but also of the higher derivatives.

Srivastava and Ramalingam (1967) and Srivastava *et al.* (1969) used a combination of techniques to evaluate the substitution pattern of hydroxyethyl starch. They carried out a periodate oxidation, reduced the resultant dialdehyde starch with borohydride, and acid hydrolysed the product. The hydrolysate was then characterized by a combination of absorption and paper chromatography.

A novel method of investigating the substitution pattern of hydroxyethyl starch has been devised by Larson (personal communication). This method avoids the formation of internal cyclic compounds which may form during acid hydrolysis of hydroxyethyl starch. Samples are dissolved in methyl sulphoxide and ethylated with ethyl iodide in the presence of sodium hydride. After hydrolysis and reaction with sodium borohydride the sample is re-ethylated. The products are then examined by gas–liquid chromatography. In samples with MS in the range 0.3–1.2, three monosubstituted glucitol derivatives, and four of the possible six disubstituted derivatives (disubstitution was not found at C6 or C3) were detected.

TABLE 1.3 *Summary of studies on the substitution pattern of hydroxyethyl starch*

Reference	Molar substitution	Relative reactivities implied from analytical data*
Husemann and Kafka (1960)	0.30–1.16	k_2 or $k_3 >> k_6$; k_P not considered
Schoch (1965)	0.90	$k_6 > k_2 > k_3$; k_P small
Lott and Brobst (1966)	0.60	$k_2 > k_3 > k_6$; k_P is finite
Srivastava and Ramalingam (1967)	0.10	$k_2 >> k_6 > k_3$; k_P not considered
Bollenback *et al.* (1969)	0.60	$k_2 > k_3 > k_6$; $k_P < k_3$
Tahan and Zilkha (1969*a, b*)	0.50–2.00	$k_P > k_2\ k_3 > k_6$
de Belder and Norrman (1969)	0.20	$k_2 > k_6 - k_3$; k_P not considered
Banks *et al.* (1971)	0.30–1.20	$k_2 : k_3 : k_6 : k_P = 15{:}1{:}1{:}10$
Yoshida *et al.* (1973)	0.55–0.94	Ratio of k_2/k_6 varied from 9.5 to 0.49, k_P of the same order as k_2 and k_6
Merkus *et al.* (1977)	0.03–1.19	$k_2 > k_6 - k_3$; k_P finite
Ozaki *et al.* (1972)	0.51	$k_2 > k_6 > k_3$

* k_2, k_3, k_6, and k_P refer to the rate constants for substitution at C2, C3, C6, and polysubstitution respectively

Results from various analytical techniques are summarized in Table 1.3. To simplify interpretation, the results have been listed in the order of reactivity of the various rate constants (see Section 1.2.3). A clear pattern emerges for, with two exceptions (Schoch 1965; Yoshida *et al.* 1973), k_2 is the rate constant with the highest value. In the two exceptions, little is known about the methods of analysis used by Schoch (1965), and in the work of Yoshida *et al.* (1973) very unusual reaction conditions were used in which the base catalysing the hydroxyethylation reaction was present in large excess (Sadota 1975). Opinion varies about the relative magnitudes of the rate constants k_3 and k_6 but there appears to be little difference between them.

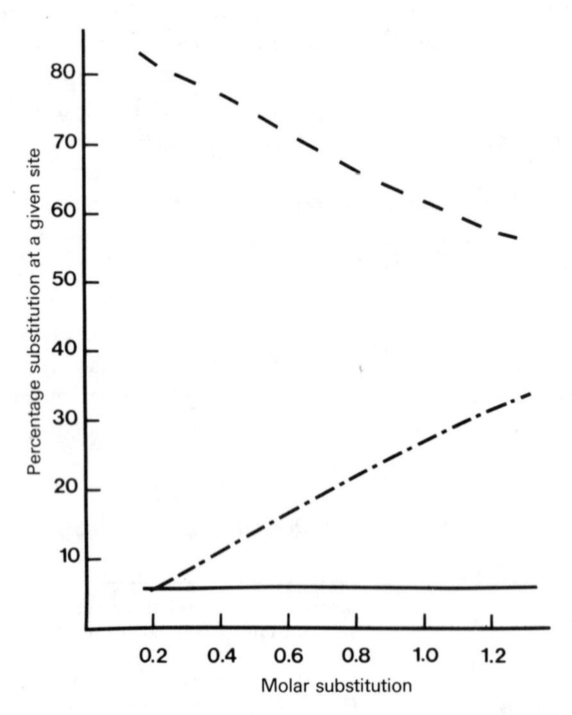

Fig. 1.6. The effect of increasing molar substitution on substitution pattern: substitution at C2 (– – –); polysubstitution (-·-·-); and substitution at C3, C6 (——).

The polymerization reaction (denoted by the rate constant k_p) is the subject of some confusion. Some reports discount polymerization completely whilst in others the overall level of substitution is so small that k_p would probably not be readily evaluated, e.g. in the samples examined by Srivastava and Ramalingam (1967) or those of de Belder and Norrman (1969). Nevertheless, there is irrefutable evidence that in widely varying reaction systems polysubstitution occurs to a significant extent (Banks *et al.* 1971, 1973; Mourits *et al.* 1976; Merkus *et al.* 1977; Yoshida *et al.* 1973). The results of Banks *et al.* (1971) best fit a reaction scheme in which $k_2:k_3:k_6:k_p$ is 15:1:1:10. The effects of increasing MS on the relative proportions of each substituent is shown in Fig. 1.6. The most significant feature is the increase in polysubstitution as MS increases (Merkus *et al.* 1977).

The effect of varying the ratios of the various rate constants – using the model presented in Section 1.2.3 – was evaluated. The substitution patterns which would be found at MS = 0.7 (typical of hydroxyethyl starch for

TABLE 1.4 *Model calculations of substitution pattern at fixed molar substitution (MS = 0.70) when the reactivities of sites for reaction vary*

Model	Ratio of rate constants*				Substitution at a named site (%)			
	k_2:	k_3:	k_6:	k_P	C2	C3	C6	CP
(a)	1	1	1	1	30	30	30	10
(b)	10	1	1	1	76	10	10	4
(c)	1	1	10	1	10	10	76	4
(d)	15	1	1	10	69	6	6	19

* k_2, k_3, k_6, and k_P are the rate constants for reaction at C2, C3, C6 or polysubstitution, respectively

pharmacological use) was calculated for varying rate constant ratios, and the results of these calculations are shown in Table 1.4. Model (d) in Table 1.4 is favoured as being closest to the most common substitution pattern found in hydroxyethyl starch but it is striking that in all models polysubstitution is significant.

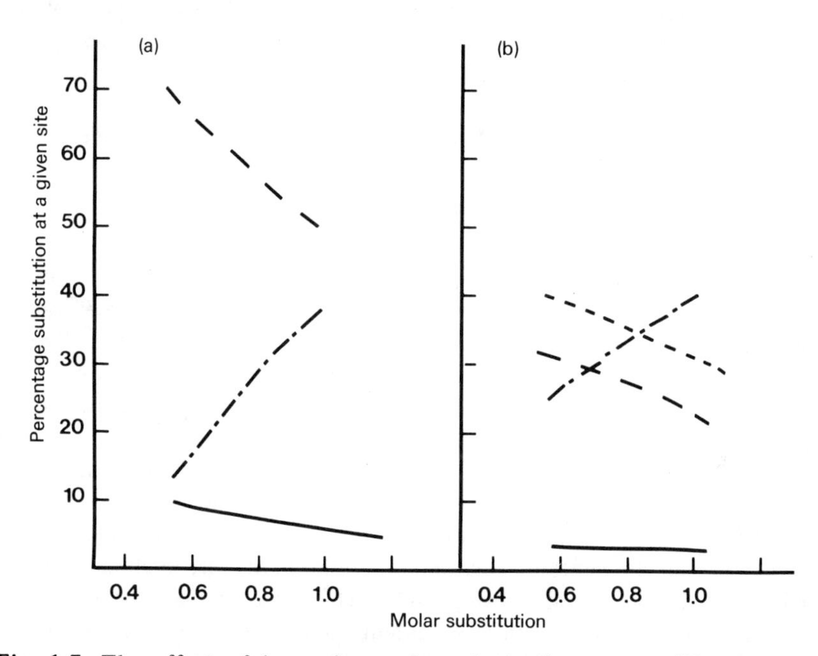

Fig. 1.7. The effect of increasing molar substitution pattern (Yoshida *et al.* 1973): substitution at C2 (----); polysubstitution (-·-·-); and substitution at C3 (——); and substitution at C6 (– – –). (a) hydroxyethyl starch substituted predominantly at C2, (b) hydroxyethyl starch substituted predominantly at C6.

When the results of Yoshida *et al.* (1973) are compared in an analogous manner to the results shown in Fig. 1.6, the resulting curves (Fig. 1.7) are strikingly similar. Although the ratio of k_2:k_6 had been reversed in these experiments (Sadota 1975), polysubstitution occurred at similar levels in both cases.

1.3.4. PREPARATION OF 2-HYDROXYETHYL ETHERS OF GLUCOSE

The hydroxyethyl ethers of D-glucose are often required for use as reference standards in the investigation of the reaction between starch and ethylene oxide. Samples of 2-O-(2-hydroxyethyl)-D-glucose, 3-O-(2-hydroxyethyl)-D-glucose, and 6-O-(2-hydroxyethyl)-D-glucose have been synthesized by allowing glucose derivatives containing suitable blocking groups to react with 2-bromoethanol, followed by removal of the blocking groups (Thewlis 1975).

1.3.5. SUMMARY

The above reports clearly show that there is considerable variation in the chemical structure of samples of hydroxyethyl starch. Nevertheless, three facts emerge. Firstly under normal conditions reaction occurs primarily at C2 on the anhydroglucose residue. Second, as MS increases – irrespective of the reaction conditions – polysubstitution becomes of increasing importance; and, third, if appropriate reaction conditions are used the relative reactivities of C2 and C6 may be changed.

1.4. THE PHYSICOCHEMICAL CHARACTERIZATION OF HYDROXYETHYL STARCH

Physicochemical characteristics of hydroxyethyl starch – the molecular weight, the molecular weight distribution, the hydrodynamic volume, and degree of hydroxyethyl group substitution – are of particular importance when hydroxyethyl starch is administered intravenously, either as a plasma volume expander or as an adjunct with previously frozen red blood cells. For example, the physical shape and size and degree of substitution of hydroxyethyl starch governs its rate of elimination from the circulatory system via the kidney and hence determines its persistence (see Section 2.2.2).

A variety of physical techniques are currently available for determining the size and shape of polymer molecules in solution, but complications arise when the polymeric material is composed of molecules of different sizes, i.e. it has a molecular weight distribution. This situation holds in the case of all native polysaccharides and in the case of hydroxyethyl starch. A first consideration, therefore, is the effect of such a molecular weight distribution on the parameter being measured, for different techniques yield different average values for a given polymer sample.

1.4.1. MOLECULAR WEIGHT AVERAGES AND THE MOLECULAR WEIGHT DISTRIBUTION

1.4.1.1. MOLECULAR WEIGHT AVERAGES

The first characteristic to be considered is the molecular weight (Greenwood and Banks 1968; Tanford 1961). When a method such as measurement of reducing power or osmotic pressure is used, the *number* of molecules in solution is effectively counted and such techniques yield a *number-average molecular weight* ($M_{\bar{n}}$). The $M_{\bar{n}}$ is given by the relation

$$M_{\bar{n}} = M_1\ (n_1/\Sigma n_i) + M_2\ (n_2/\Sigma n_i) + M_3\ (n_3/\Sigma n_i) + \ldots nM_i\ (n_i/\Sigma n_i) \quad (1.19)$$

where n_1 is the number of molecules of molecular weight M_1, n_2 is the number of molecules of molecular weight M_2, etc. The summation (Σn_i) is the total number of molecules present and hence each term ($n_i/\Sigma n_i$) is the mole fraction of each species. Hence,

$$M_{\bar{n}} = n_i M_i/\Sigma n_i. \quad (1.20)$$

However, as solutions are generally prepared by weight, $M_{\bar{n}}$ can be expressed in the form:

$$M_{\bar{n}} = \frac{1/\Sigma w_i}{M_i} \quad (1.21)$$

where w_i is the weight fraction of the species of molecular weight M_i.

In contrast, in methods such as light-scattering, the measured property is a function of the *mass* of the polymer and the resultant molecular weight is the *weight-average value* ($M_{\bar{w}}$), where

$$M_{\bar{w}} = M_1\ (nM/\Sigma n_i m_i) + M_2\ (n_2 M_2/\Sigma n_i M) + \ldots M_i\ (n_i M_i/\Sigma n_i M_i)$$

$$= \Sigma n_i M_i^2/\Sigma n_i M_i \quad (1.22)$$

$$= \Sigma w_i\ M_i. \quad (1.23)$$

In these relations, the factor ($n_i M_i/\Sigma n_i M_i$) is the *weight-fraction* of each species.

The difference between $M_{\bar{n}}$ and $M_{\bar{w}}$ is most easily illustrated by a numerical example. For instance, the mixing of equal *numbers* of molecules of molecular weight of 10 000 and 100 000 gives $M_{\bar{n}}$ = 55 000 and $M_{\bar{w}}$ = 91 800. The mixing of equal *weights* of the same species gives $M_{\bar{n}}$ = 18 200 and $M_{\bar{w}}$ = 55 000. Clearly, $M_{\bar{n}}$ is the simple arithmetic mean value, whilst $M_{\bar{w}}$ is influenced predominantly by the higher molecular weight species present in the mixture.

There are other molecular weight averages but these are not effectively significant in the context of hydroxyethyl starch. It might be mentioned, for example, that sedimentation and diffusion measurements or Archibald sedimentation measurements – all methods used in early studies of hydroxyethyl starch – yield a complex average which is not simply related either to $M_{\bar{n}}$ or $M_{\bar{w}}$.

Viscosity is another technique widely used for the characterization of polymers but it is not an absolute method and hence such measurements have to be calibrated. Nevertheless, the limiting viscosity number [η] is a very convenient, accurate, and easily measured characteristic which is often used as an indicator of polymer size and shape. $[\eta]$ is obtained from the relation

$$\eta_{sp}/c = [\eta] + k_i [\eta]^2 c \qquad (1.24)$$

where η_{sp} is the specific viscosity, c is the concentration of polymer in solution, and k^i is a constant.

1.4.1.2. MOLECULAR WEIGHT DISTRIBUTION

A typical molecular weight distribution curve for a polymer is shown in Fig. 1.8. The relative positions of the values of $M_{\bar{n}}$ and $M_{\bar{w}}$ are also shown. These two averages diverge as the spread of the molecular weight distribution increases. The closer the value of the ratio $M_{\bar{w}}/M_{\bar{n}}$ approaches unity the narrower the molecular weight distribution. Samples of hydroxyethyl starch have been characterized by such techniques (see Table 2.1).

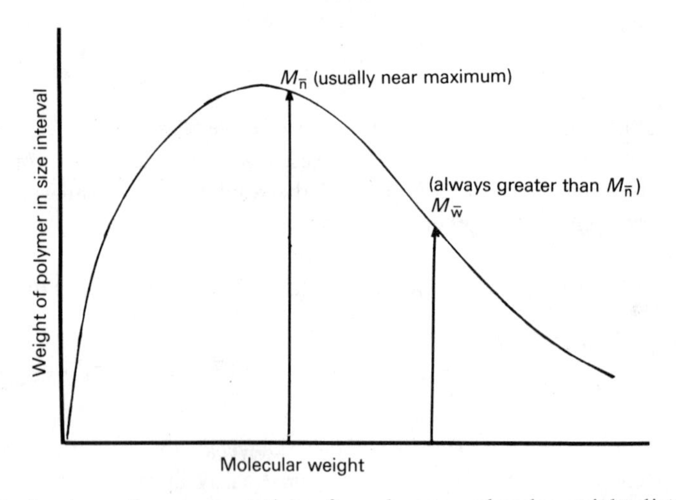

Fig. 1.8. A schematic representation of a polymer molecular weight distribution.

For complete characterization, however, the molecular weight distribution should be determined experimentally. Methods of achieving this estimation depend on separating the polymer into discrete fractions of narrow molecular weight distribution, e.g. fractionating the polymer. The amount and molecular weight of each fraction can then be determined. The resultant graph of cumulative amount against molecular weight is the integral molecular weight distribution (Fig. 1.9), which may be differentiated to yield the differential molecular weight distribution curve shown in Fig. 1.10.

Fractionation can be either stepwise or continuous. The first technique uses fractional precipitation, in which the polymer is dissolved in a good solvent and then non-solvent is gradually added. Because the solubility of most polymers is

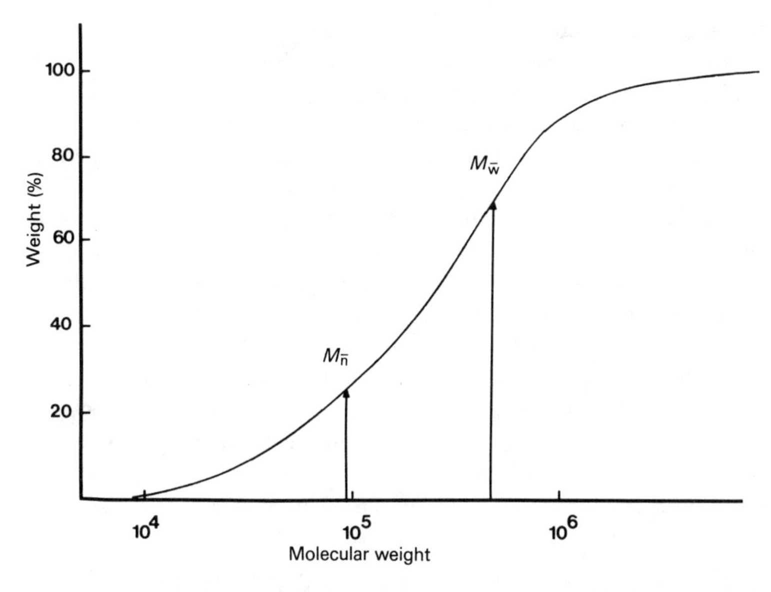

Fig. 1.9. The integral molecular weight distribution of hydroxyethyl starch (McGaw Laboratories sample code S3184A).

inversely related to molecular weight, the first precipitate contains the largest polymers, the next fraction contains the next largest, and so on. Each precipitate is weighed and its molecular weight ($M_{\bar{n}}$ or $M_{\bar{w}}$) is determined to obtain the distribution curve discussed above. This process is tedious and time consuming but was used in all early studies of hydroxyethyl starch.

A major advance in the study of the molecular weight distribution was achieved by the development of the technique of molecular exclusion filtration (MEF).

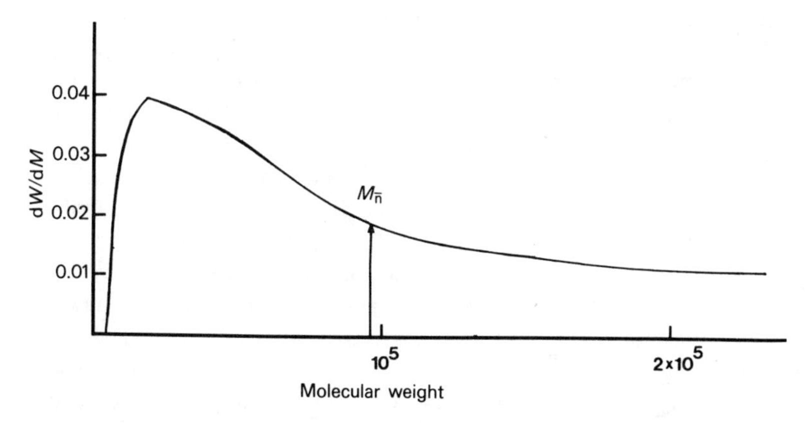

Fig. 1.10. The differential molecular weight distribution of hydroxyethyl starch (McGaw Laboratories sample code S3184A).

In this procedure the polymer sample is carried by a moving liquid phase through a bed of swollen gel particles. The porosity of the gel particles is carefully controlled. During passage through the column the volume available to any given macromolecule depends on its size; small molecules may penetrate the gel phase readily whilst large molecules are completely excluded into the volume outside the gel (see Appendix 2 for details). Thus when a mixture of polymers of varying size is eluted through such a column, the large molecules are eluted first and the smallest molecules last. Thus the mixture is separated on the basis of molecular size and an analysis of the polymer concentration in the eluate from the column can be correlated with the molecular weight distribution (for an example of the technique applied to dextran see Granath and Kvist 1967).

In practice, elution volumes for a given gel column are calibrated against a range of well-defined polymer fractions for which $M_{\bar{n}}$ and $M_{\bar{w}}$ are known. The calibration curve is then inserted into an appropriate computer program (Nilsson and Nilsson 1974). Consequently, elution profiles of unknown polymer samples may be converted into molecular weight distribution curves.

1.4.2. THE RELATION BETWEEN VISCOSITY AND MOLECULAR WEIGHT

The hydrodynamic properties of a polymer – its shape in solution and its interaction with solvent – can often be obtained from measurements of $[\eta]$ in combination with those of $M_{\bar{n}}$ and $M_{\bar{w}}$. The appropriate theory and its application to starch have been extensively reviewed elsewhere (Banks and Greenwood 1975; Tanford 1961). A significant amount of information can be obtained by examination of the Mark-Houwink relations

$$[\eta] = K_a M^a \tag{1.25}$$

$$S_0 = K_b M^b \tag{1.26}$$

where $[\eta]$ is the limiting viscosity number, S_0 is the sedimentation coefficient at infinite dilution, M is the molecular weight, and K_a and K_b are constants for given polymer-solvent systems. The exponents a and b in eqns 1.25 and 1.26 have values which depend on: (i) the degree of branching of the polymer; (ii) the shape of the macromolecules in solution (for example, a flexible Gaussian coil or an extended rod) and (iii) the extent of polymer-solvent interaction.

Measurements made in a *Flory*-theta-solvent are of particular value for, in this special case, solvent-polymer interaction is minimized and only the first two factors described above influence the value of a in eqn 1.25. As shown in Table 1.5, certain deductions about polymer shape may be drawn from the value of a obtained in a theta-solvent. The values shown in Table 1.5 refer to measurements made in a theta-solvent; and, although a theory has been proposed for systems in which considerable polymer-solvent interaction occurs, it has limited practical application (Banks and Greenwood 1975; Banks *et al.* 1971).

TABLE 1.5 *Interpretation of the Mark–Houwink relation in a theta-solvent*

Mark–Houwink exponent	Polymer shape
$a < 0.5$	Branched molecule
$a = 0.5$	Flexible, Gaussian coil
$a > 0.5$	Having an extended or rigid backbone

1.4.3. PHYSICOCHEMICAL STUDIES ON HYDROXYETHYL STARCH

Comparatively few investigations of the physicochemical characteristics of hydroxyethyl starch have been made. The earliest report by Husemann and Resz (1956) dealt with samples of amylose which had been derivatised to various extents. These authors fractionated the polymer by acetone precipitation, measured the $M_{\bar{n}}$ of the resultant fractions by osmotic pressure, and then related $M_{\bar{n}}$ to the limiting vicosity number $[\eta]$. Their data yield the values of a shown in Table 1.6. The results are unusual for the extrapolation of the values of a to MS = O, that is native amylose, because it gives a value very much greater than unity when, in fact, $a = 0.50$ for native amylose in a theta-solvent (Banks and Greenwood 1975).

TABLE 1.6 *The effect of molar substitution on the properties of hydroxyethyl amylose**

Molar substitution	a
0.31	1.65
0.61	1.21
1.08	0.64

* Data recalculated from Husemann and Resz (1956)

The first widely reported physicochemical study of clinical hydroxyethyl starch was by Greenwood and Hourston (1967). These authors made an overall comparison of the properties of five commerical samples, with results expressed for $[\eta]$ and $M_{\bar{w}}$ (Table 1.7). The molecular weight was obtained by light-scattering measurements but sedimentation measurements in the analytical ultracentrifuge were also carried out. These sedimentation studies indicated that a very wide molecular weight distribution was present in all samples.

TABLE 1.7 *Physicochemical properties of clinical hydroxyethyl starch*

Samples*	MS	$[\eta]$	$M_{\bar{w}}$
1	0.83	0.26	320 000
2	0.92	0.23	310 000
3	0.93	0.21	230 000
4	0.93	0.18	184 000
5	0.93	0.14	106 000

*From McGaw Laboratories, Irvine, California

Hydroxyethyl starch was also studied by Cerny *et al.* (1967). A sample of hydroxyethyl starch (MS = 0.85) was fractionated by acetone and iso-propanol precipitation. After measuring the $M_{\bar{n}}$ by osmotic pressure and the $M_{\bar{w}}$ by light scattering, the following relations with viscosity were obtained:

$$[\eta] = 5.29 \times 10^{-3} \cdot M_{\bar{w}}^{0.30} \tag{1.27}$$

$$[\eta] = 3.77 \times 10^{-3} \cdot M_{\bar{n}}^{0.35} \tag{1.28}$$

The difference between equations (1.27) and (1.28) results from the variation in the distribution of molecular weight between the fractions. The measurements were not carried out in a theta-solvent for, in both light-scattering and osmotic-pressure measurements, positive values were found for the viral coefficients. To overcome this difficulty, Cerny and his colleagues (1967) attempted to apply theoretical treatments to give an estimate of solvent–polymer interaction. However, they found the theory inapplicable to hydroxyethyl starch. Thus the only firm conclusions to be drawn from the values of the exponents in equations (1.27) and (1.28) are that they reflect the branched nature of the polymer.

Extensive experiments were reported by Granath *et al.* (1969) for clinical hydroxyethyl starch of varying levels of substitution. These authors fractionated hydroxyethyl starch by preparative MEF and measured $M_{\bar{n}}$ by osmotic pressure and $M_{\bar{w}}$ by light scattering for all fractions. The relation

$$[\eta] = 2.91 \times 10^{-3} \cdot M_{\bar{w}}^{0.35} \tag{1.29}$$

was obtained, and the low value of a = 0.35 was taken to indicate that the polysaccharides were highly branched (Sakamoto *et al.* 1977). As MS increased, Granath *et al.* (1969) noted a concomitant increase in viscosity for a given molecular weight, and they suggested that this phenomenon was a result of hindered rotation about the glycosidic bond in the substituted polymer. The technique of MEF was applied to the samples and estimates of the molecular weight distribution were obtained from the ratios of $M_{\bar{w}}/M_{\bar{n}}$ (See Table 1.8). The results in Table 1.8 show the usefulness of MEF for physicochemical characterization of hydroxyethyl starch and also indicate the very broad molecular weight distribution in the clinical samples which were examined.

TABLE 1.8 *Molecular weight data for some hydroxyethyl starch samples**

Sample	MS	Calculated values from MEF			Measured values	
		$M_{\bar{w}}$	$M_{\bar{n}}$	$M_{\bar{w}}/M_{\bar{n}}$	$M_{\bar{w}}$	$M_{\bar{n}}$
1	0.72	525 000	72 000	7.3	540 000	–
2	0.52	125 000	37 080	3.3	131 000	–
3	0.66	144 000	33 700	4.3	143 000	38 000
4	0.59	106 000	25 000	4.3	110 000	–
5	0.65	141 000	17 000	8.3	141 000	–

* From Granath *et al.* (1969)

In a more recent study, these authors (Granath and de Belder, personal communication) used the same techniques to obtain the molecular weight distribution of a clinical sample of hydroxyethyl starch (Fig. 1.4 and 1.10): the corresponding molecular weight values were $M_{\bar{w}} = 488\,100$, $M_{\bar{n}} = 93\,400$ and $M_{\bar{w}}/M_{\bar{n}} = 5.23$.

The viscosity-molecular weight relation for hydroxyethyl starch was reported by Tamada *et al.* (1971) to be:

$$[\eta] = 4.72 \times 10^{-6} \cdot M^{0.52} \tag{1.30}$$

The result is difficult to reconcile with similar data from Cerny *et al.* (1967) or Granath *et al.* (1969) perhaps because Tamada and his colleagues used a very different experimental technique: molecular weights were estimated using the Archibald method in the ultracentrifuge.

All of these physicochemical studies show that hydroxyethyl starch has a comparatively low viscosity in relation to its molecular weight. For example, Greenwood and Hourston (1967) made the comparison shown in Table 1.9.

TABLE 1.9 *A comparison of the molecular properties of hydroxyethyl starch and dextran*

Hydroxyethyl starch		Dextran	
$[\eta]$	$M_{\bar{w}}$	$[\eta]$	$M_{\bar{w}}$
0.27	320 000	0.25	65 600
0.23	310 000	0.25	65 600
0.21	230 000	0.21	50 000
0.18	184 000	0.18	31 400
0.13	106 000	0.13	18 400

For the same viscosity potential, hydroxyethyl starch is about five times larger in molecular weight than clinical dextran. Consequently, for a given molecular size, molecular weight distribution will be less important for hydroxyethyl starch than for dextran. In addition, the osmotic pressure of comparable solutions of hydroxyethyl starch will be *less*, as osmotic pressure is inversely related to molecular size.

The interaction of hydroxyethyl starch and albumin has been examined by means of osmotic pressure effects and the hydroxyethyl starch has been shown to be as effective a plasma volume expander as dextran (Cerny *et al.* 1968). A viscosity study was carried out on hydroxyethyl starch/erythrocyte mixtures; but, due to the heterogeneous nature of hydroxyethyl starch, the authors were reluctant to define the ideal size and shape of hydroxyethyl starch for use as a volume expander.

1.4.4. A COMPARISON WITH CLINICAL DEXTRANS

1.4.4.1. THE CHEMICAL STRUCTURE OF DEXTRAN

Dextrans are a group of α-glucans produced by different bacteria, the most important being *Leuconostoc mesenteroides* and *Leuconostoc dextranicum* (see Neely (1960) for a review of the characteristics of different dextran-producing organisms). These bacteria use sucrose to produce dextran by the following reaction:

$$n \text{ sucrose} \longrightarrow \underset{\text{(dextran)}}{\text{polyglucose}} \longrightarrow + n \text{ fructose} \tag{1.31}$$

Methylation studies have shown that the main glycosidic bond in all dextrans is the α-1:6-linkage but that α-1:3-, α-1:2-, and α-1:4- links may be present, depending on the strain of organism used. In fact, different strains of the same bacterium may produce dextrans of differing structure. Evidence for the nature of the branch-points has been obtained from periodate oxidation, infra-red absorption spectra, and optical rotation measurements (Neely 1960).

As with hydroxyethyl starch, the physicochemical properties of a dextran depend on the molecular weight and the type and extent of branching. Molecular weight studies of *native* dextrans have shown these polysaccharides to be extremely large (e.g. degree of polymerization (DP) $>10^6$) and they often exist an aggregates (Burchard and Cowie 1972).

Acid-degraded dextrans from *L. mesenteroides* have been used widely as plasma volume expanders, the objective of the hydrolysis being to reduce the polysaccharide to a molecular size comparable with that of the plasma proteins, i.e. $M_{\bar{w}} = 75\,000$.

Most clinical dextran is currently produced by a single company in Sweden (Pharmacia Fine Chemicals). Because dextran can vary so much in type and extent of branching, commerical production is restricted to polymer from a single strain of *L. mesenteroides*. This strain (B512) was first isolated and studied by Jeanes and her co-workers (for a review see Neely 1960). In this species of dextran over 90 per cent of the linkages are of the α-1:6-type in the main and side-chains, whilst the remainder are α-1:3-branch points (Fig. 1.11). The exact length of the side chains is not established but the most current evidence suggests that they may vary in size from single glucose units to chains of many residues.

1.4.4.2. PHYSICOCHEMICAL STUDIES OF CLINICAL DEXTRAN

Acid-degraded dextran from strain B512 *L. mesenteroides* has been the subject of various physicochemical studies to obtain the relation between molecular weight and viscosity. Wales *et al.* (1953) measured the molecular weight of dextran samples by sedimentation equilibrium and found the relation:

$$[\eta] = 10^{-3} \cdot M^{0.5} \tag{1.32}$$

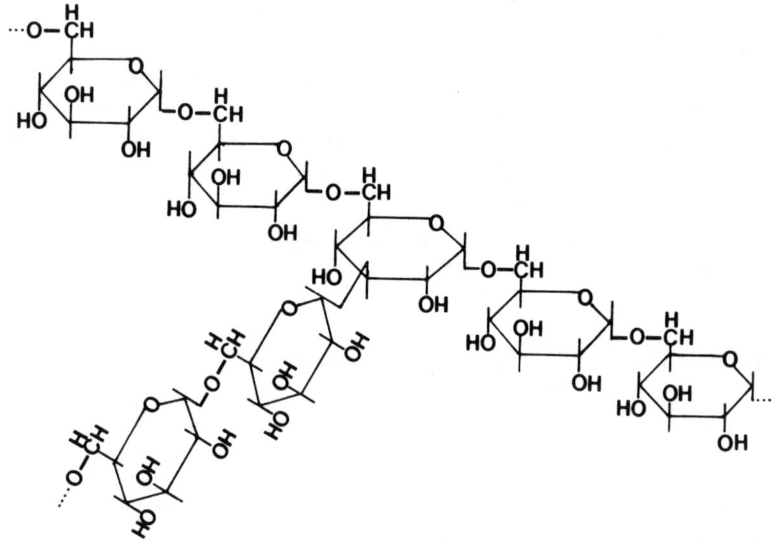

Fig. 1.11. The structure of dextran from *L. mesenteroides* B512.

for aqueous solutions. The data obtained by Riddick *et al.* (1954) from light-scattering measurements on dextran fractions also fitted this equation.

Viscosity, sedimentation, and light-scattering properties of clinical dextran were also measured by Senti *et al.* (1955), who also found that equation (1.32) held for molecular weights of under 100 000. Above this molecular weight, the value of a in the Mark–Houwink equation decreased rapidly, as would be expected from the branched nature of this polysaccharide. Indeed, when Granath (1958) also attempted to correlate the degree of branching of this clinical dextran with hydrodynamic behaviour, she found the reaction:

$$[\eta] = 2.43 \times 10^{-3} \cdot M_{\bar{W}}{}^{0.42} \qquad (1.33)$$

to hold.

In these studies, the low value of a in the Mark–Houwink equation reflects the branched nature of the polysaccharide.

It is significant, however, that the viscosity of dextran is considerably higher than that of hydroxyethyl starch of comparable molecular weight (see Table 1.9). Consequently, for clinical use both the average molecular weight and the molecular weight distribution of the dextran are of considerable importance and have to be carefully controlled. Adverse clinical effects occur for dextran molecules whose molecular weight approaches 100 000.

1.4.4.3. PROPERTIES OF CLINICAL DEXTRAN

The molecular weight distribution of commercial hydrolysed dextran is controlled by repeated fractionation to remove the very small and the very large polysaccharide molecules. The molecular weight distributions of products from Pharmacia Fine Chemicals are listed in Table 1.10. As discussed earlier in Section 4.1.2., a measure of the spread of molecular weight in a polymer sample is obtained from the ratio $M_{\bar{w}}/M_{\bar{n}}$. With clinical dextran samples, $M_{\bar{n}}$ can be obtained by end-group analysis, e.g. by measuring the number of free potential reducing groups at the end of each molecule. This estimation can be made by use of a colorimetric reagent such as dinitro-salycilic acid, or modified Somogyi reagent (Senti *et al.* 1955). The $M_{\bar{w}}$ is obtained directly from light-scattering measurements.

TABLE 1.10 *Molecular weight distribution for dextran*

Species	$M_{\bar{w}}$	40% to limit
dextran 40	40 000	15 000–75 000
dextran 70	70 000	20 000–115 000

The application of MEF (as described earlier) to clinical dextran has enabled the molecular weight distribution to be evaluated routinely (Alsop *et al.* 1977; Granath and Kvist 1967).

The very wide difference in the spread of molecular weights in clinical dextran and in hydroxyethyl starch is shown in Fig. 1.12.

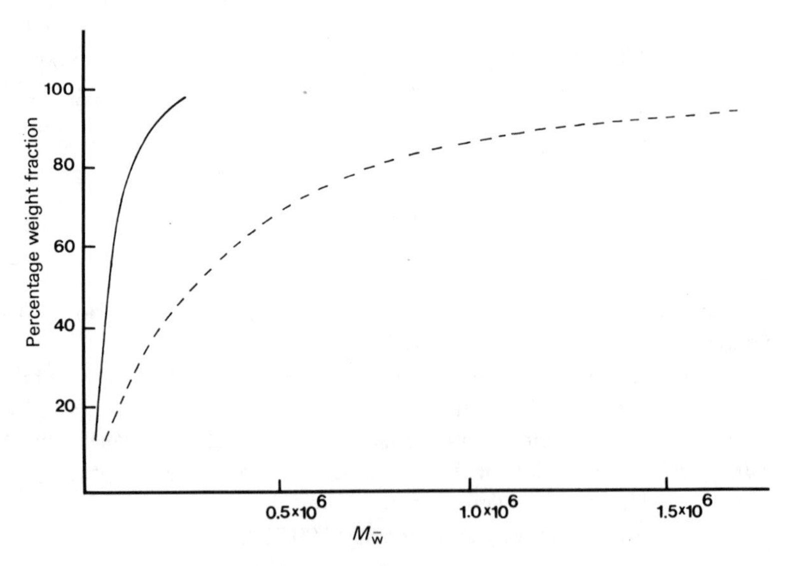

Fig. 1.12. A comparison of the molecular weight distribution curves for clinical hydroxyethyl starch and dextran. Dextran 70 (——) and hydroxyethyl starch (– – –) (McGaw Laboratories sample code S3234A).

The osmotic pressure of dextran 70 (6 per cent solution) is about 800 mm of water, whilst that for dextran 40 (10 per cent solution) is about 2300 mm of water. *In vitro*, therefore, both solutions exert a considerably higher osmotic pressure than normal plasma (7.5 per cent protein solution), which is about 350 mm of water.

The effect of a shear on solutions of dextran 40 mixed with blood and plasma has been reported by Groth (1966).

1.5. THE ENZYMIC DEGRADATION OF HYDROXYETHYL STARCH

1.5.1. THE EFFECT OF MS AND DS ON THE ENZYMIC DEGRADATION OF HYDROXYETHYL STARCH

Two distinct types of starch-degrading enzymes exist: those whose action pattern proceeds stepwise from a non-reducing end of a starch chain – the *exo-amylases* – and those enzymes capable of hydrolysing internal glycosidic linkages within the chains – the *endo-amylases* (French 1973, 1975; Greenwood 1968; Greenwood and Milne 1968). The introduction of chemical substituents into the starch has a general inhibitory effect on all processes of enzymic degradation; but the effect is more pronounced in the case of the exo-amylases, such as β-amylase or phosphorylase, than with the typical endo-enzyme, α-amylase. The reason for this behaviour is shown schematically in Fig. 1.13 for a stylized linear chain derivatized to a molar substitution of 0.15. A chain of DP of 40 is shown, and it can be seen that the exo-enzymes may remove successively five glucose residues at most from the non-reducing end before it meets a modified residue which cannot be removed. The DP is thus reduced to 35. Exo-amylase activity is thus of comparatively little significance in the hydrolyis of hydroxyethyl starch. However, the endo-amylase may cleave the chain between substituent groups, and consequently the polymer may be almost completely hydrolysed. This fact has been recognized for a long time and was used by Ziese (1934, 1935) and Scholander and Myrback (1951) as a means of differentiation of endo- from exo-amylase activity. Greenwood and Hourston (1967) confirmed the validity of these results for a commercial sample of hydroxyethyl starch prepared for the purpose of plasma volume expansion. This sample was essentially unaffected by the action of β-amylase acting alone or in the presence of weak α-amylase (z-enzyme).

Physiologically, α-amylase is the most important enzyme for the degradation of hydroxyethyl starch. The rate of α-amylolysis of hydroxyethyl starch has been studied by several workers. Husemann and Resz (1956) examined the action of α-amylase from *Aspergillus oryzae* on the rate of degradation of hydroxyethyl amylose of varying molar substitutions. The rate of hydrolysis fell markedly with increasing MS but was still finite at MS = 1.03. As this level of MS implies complete substitution of the starch – that is, every anhydroglucose residue has one substituent group – either substitution was not uniform or hydroxyethylation only slowed down, not prohibiting α-amylolysis.

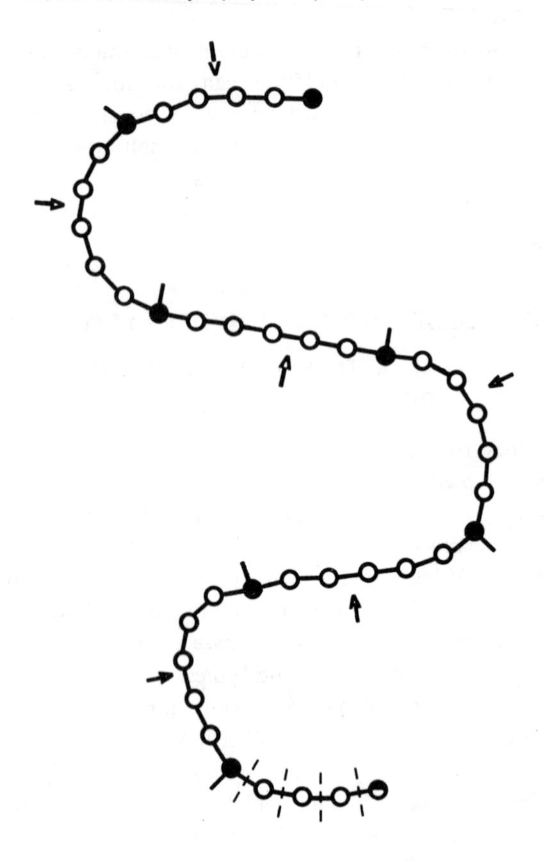

Fig. 1.13. A model for enzyme attack on hydroxyethyl starch: substituted glucose (♦); unsubstituted glucose (○); non-reducing chain end (◑); reducing chain end (●); bonds susceptible to stepwise action of an exo-enzyme (¦); and areas of polymer susceptible to random cleavage by an endo-enzyme (↑).

Evidence from the work of Yoshida *et al.* (1973) favours the earlier proposal that substitution does not occur uniformly in hydroxyethyl starch and polyethylene oxide side-chains are formed as hydroxyethylation proceeds (Banks *et al.* 1973). (This reaction and its consequences have been dealt with in Section 1.3.) Therefore, measurement of MS gives little indication of the actual susceptibility of hydroxyethyl starch to attack by α-amylase. However, as shown in Fig. 1.14, the extent of hydrolysis can be meaningfully related to DS. The increase in resistance to enzymic attack is not linear, but increases steeply above DS = 0.6. Exactly the same conclusion may be drawn from the results of Yoshida *et al.* (1973), although a different assay system was used to evaluate the extent of polymer degradation. These data show that DS gives a true indication of the likely persistence of hydroxyethyl starch *in vivo*.

The data of Husemann and Resz (1956) may be best interpreted in terms of a model in which MS gives a misleading index of susceptibility to enzyme attack.

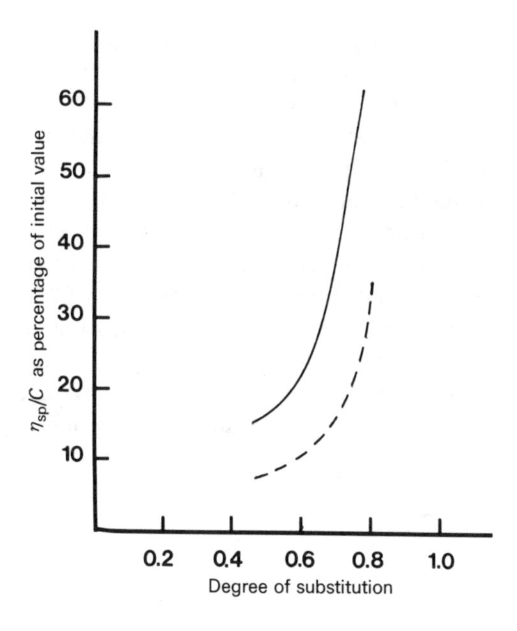

Fig. 1.14. The effect of α-amylase on hydroxyethyl starch. The experimental conditions were those of Banks *et al.* (1973). Time of exposure to enzyme (min): 2 (——) and 1500 (– – –).

There is a suggestion from the work of several investigators that not all types of substitution will have equal effects (Chan 1975; French 1972; Robyt and French 1970). These investigators studied the hydrolysis products formed by the action of porcine pancreatic α-amylase acting on O-(2-hydroxyethyl)-amylose. A detailed analysis of the hydrolysis products was then related to the nature of the active site of the enzyme (see Fig. 2.2, p. 33). On the basis of an hypothesis which included the five-unit glucose binding site of α-amylase, it was proposed that 6-O substituents are permitted on the glucose residues at subsites I, II, and V (the reducing end is numbered I) but prohibited at III and IV. 2-O-substitution is permitted at I and IV but prohibited at III. 3-O-substitution is permitted at II, IV, and V but prohibited at III (Chan 1975; French *et al.* 1974). These data are consistent with a five-glucose binding site, in which the catalytic groups are located at bond 2 (Robyt and French 1970).

1.5.2. THE EFFECT OF SUBSTITUTION PATTERN ON α-AMYLOLYSIS OF HYDROXYETHYL STARCH

Matsushima and co-workers (Arita *et al.* 1970; Arita and Matsushima 1970; Fujinaga *et al.* 1968) studied the substrate specificity of α-amylase (Taka-amylase A) on synthetic preparations of O-methylated amylose with varying MS and on the mono-O-methyl derivatives of phenyl-α-maltoside and have shown that hydrolysis is either inhibited or significantly slowed down when substitution takes place at C2, whilst substitution at C6 is much less effective.

With hydroxyethyl starch, an exactly analogous effect was reported by Yoshida *et al.* (1973). These authors found that when the ratio of substitution at C2:C6 was greater than 2, degradation was largely unaffected by substitution pattern. However, as the ratio of substitution at C2:C6 decreased below 2, the rate of degradation increased. For example, Yoshida *et al.* (1973) measured reducing power after degradation for a series of samples of hydroxyethyl starch in which MS was constant but the ratio of substition at C2 and C6 varied. As shown in fig. 1.15, when the substitution ratio C2:C6 was 0.5, the extent of hydrolysis was three times greater than when the appropriate ratio exceeded a value of 2. This result is corroborated by the later work of Chan (1975) and is clearly a direct consequence of the substrate specificity of α-amylase.

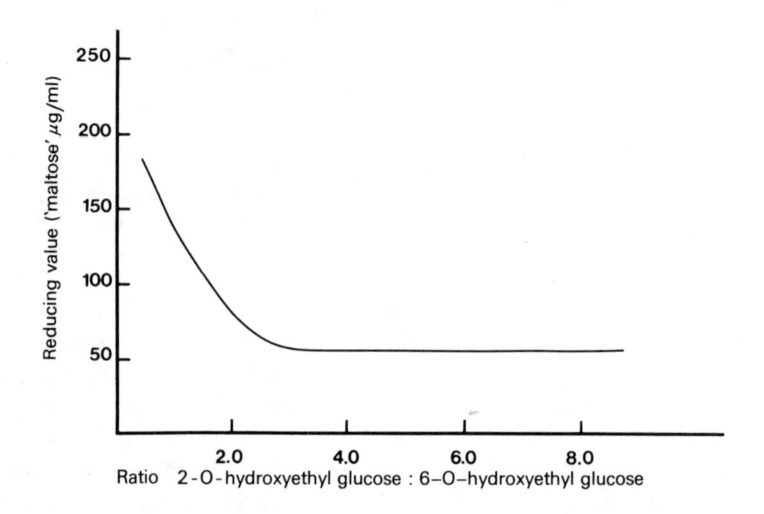

Fig. 1.15. The relation between substitution pattern and degradation of hydroxyethyl starch. (From Yoshida *et al.* 1973.)

1.6. PHYSICAL STABILITY OF HYDROXYETHYL STARCH IN SOLUTION

An important practical problem in the storage of volaemic colloids for use in disasters is the stability of solutions. Both dextran 40 and dextran 70 precipitate on storage and this is hastened by temperature fluctuations (Ewald *et al.* 1964; Lee-Benner and Walton 1965). Solutions of hydroxyethyl starch, however, are quite stable under these conditions, probably because of the compact branched molecular structure (Lee-Benner and Walton 1965; Shields *et al.* 1965).

2. Catabolism, excretion, and tissue storage of hydroxyethyl starch

2.1. INTRODUCTION

As discussed in Section 1.5, hydroxyethyl group substitution has a general inhibitory effect on the enzymic degradation of hydroxyethyl starch, but this effect is more pronounced for the *exo*-amylases rather than for the typical *endo*-amylase, α-amylase. The inhibitory effect of substitution by hydroxyethyl groups depends on two variables, *amount* and *pattern* of substitution. Superimposed on these two variables are the rather heterogeneous populations of various molecular-weight polymers contained in the injected solution of hydroxyethyl starch (see Figs. 1.8 and 1.9). In attempting to construct a model of the catabolism and subsequent elimination of hydroxyethyl starch, it is necessary to envision a series of dynamic and ever-changing variables and their relation to each other with time.

For the purposes of this discussion, it may be useful to define the contribution of each variable, independently and without regard to time. As clearly pointed out by Merkus *et al.* (1977), the *amount* of hydroxyethyl group substitution influences the number of attachments of these groups to individual glucose residues contained in the starch polymer. As the MS becomes higher, the proportion of mono-substituted residues of glucose decreases, while the number of di-, tri-, and tetra-substituted residues increases (Fig. 2.1). This observation might appear to be inconsequential if it were not for the detailed experiments performed by Chan (1975). In his studies, Chan was able to predict what *patterns* of hydroxyethyl group substitution on individual glucose residues contained in the five-unit substrate would be susceptible to attack by α-amylase (Fig. 2.2) (see Section 1.5). His work has indicated that specific substitutions can be incorporated into the substrate unit without hindering hydrolysis by α-amylase. Having taken these two variables out of context, I will attempt to merge them into a coherent, plausible theory by introducing a third variable.

As described in Section 1.3.3 substitution of the hydroxyethyl group is favoured most often at the C2 of the glucose ring rather than the C3 or the C6 (Fig. 2.1). Thus, logical deduction would lead to the conclusion that, if the *amount* of hydroxyethyl group substitution were increased, enchancing multiple attachments to individual glucose residues in the substrate unit, the chances of unfavourable *patterns* would be increased and attack by α-amylase would be reduced accordingly. Before attempting to support this hypothesis, primary consideration will be given to the various populations of molecular weights of polymers contained in a solution of soluble *unsubstituted* starch. Physicochemically the

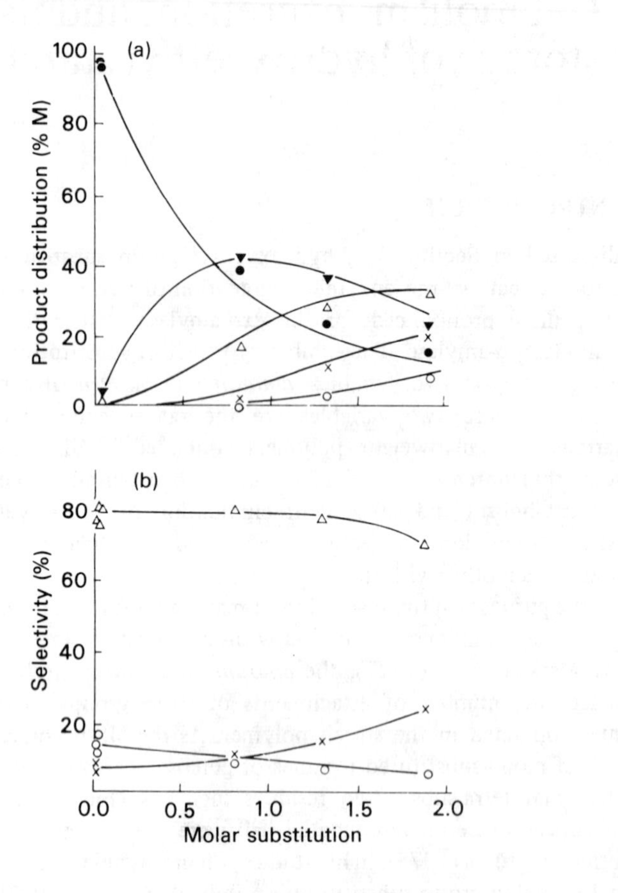

Fig. 2.2. *In vitro* studies conducted by Chan (1975) have shown that within a As the molar substitution is increased, the proportion of di- (△), tri- (X), and tetra- (○) substituted glucose residues increases, while free glucose (●) and the number of mono- (▼) substituted glucose residues decreases. (B) Selectivity of the mono-substitution reaction at C2 (△), C6 (X), and C3 (○) at various degrees of molar substitution. (From Merkus *et al.* 1977.)

molecular weight polymers of various weight contained in the solution can be defined mathematically by the $M_{\bar{n}}$ and $M_{\bar{w}}$ (see Section 1.4).

Following intravenous injection of soluble, *unsubstituted* starch, the quantity of this material voided into the urine initially could be approximated by determining the $M_{\bar{n}}$ of the infused polymer population. By knowing the limit of the renal threshold and the number of molecules whose weight constituted 50 per cent of the injected polymers, a prediction can be made of how many molecules would be filtered 1-2 hours after injection. For example, if the $M_{\bar{n}}$ of the injected solution of starch was approximately that of the limit of the renal threshold, a large proportion of the injected mass would then be eliminated

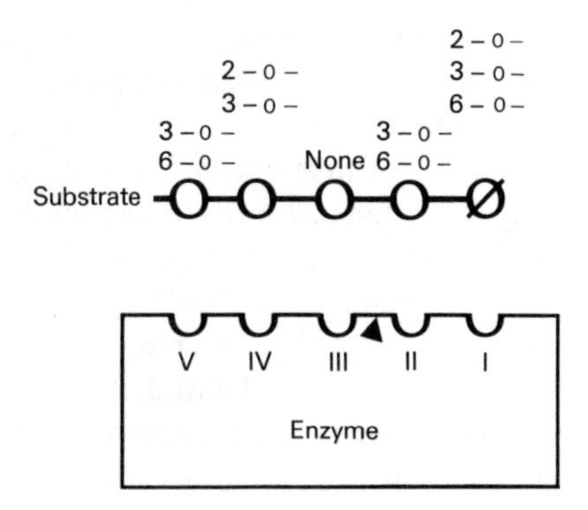

Fig. 2.2. *In vitro* studies conducted by Chan (1975) have shown that within a specific five-unit amylose substrate (○ = glucose monomer, ∅ = reducing end) contained within the amylopectin polymer, acceptable patterns of hydroxyethyl group substitution do exist. The active site (▲) of α-amylase is shown relative to the five-unit substrate (see Section 1.5 for details).

by simple filtration. Under these conditions, the $M_{\overline{n}}$ could be considered a *constant*; its value is a property of the starting material rather than the end-result of hydrolysis. The $M_{\overline{w}}$, in contrast, is a variable; it describes a population of molecules by molecular weight. As the $M_{\overline{w}}$ is increased, elimination of the injected mass will be dependent on the action of α-amylase.

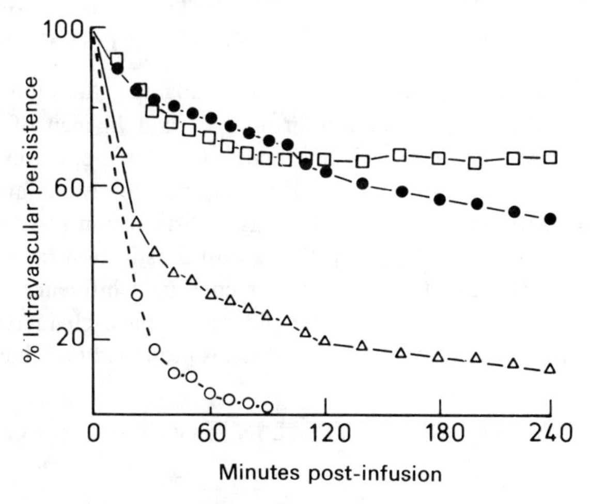

Fig. 2.3. Progressive blood concentrations after infusions (1.8 g/kg) of either dextran (□), hydroxyethyl starch (●), glucose (△), or soluble starch (○) to dogs from which the same volume of blood had been withdrawn. (From Thompson *et al.* 1962.)

Superimposed on the variable $M_{\overline{w}}$ is the inherent distribution of α-amylase-resistant regions contained in the highly branched structure of starch (Brammer *et al.* 1972). This is exemplified by the exponential rather than first-order description of the elimination of *unsubstituted* starch from blood (Fig. 2.3). Considering now hydroxyethylation in relation to $M_{\overline{n}}$ and $M_{\overline{w}}$, the corresponding catabolism becomes highly complex. For the sake of discussion, the $M_{\overline{n}}$ can be considered to describe a population of substituted polymers that would be little affected by the *amount* or *pattern* of hydroxyethylation; these polymers would generally be below the limit of the renal threshold and would be eliminated because of their low molecular weight. The remaining polymers of substituted starch would be acted on by α-amylase in accordance with the *amount* and *pattern* of substitution incorporated into the parent molecule. It is this latter instance that makes prediction of elimination very difficult.

In addition to the *amount* and *pattern* of hydroxyethylation, consideration must also be given to the fact that portions of the starch molecule will be more substituted than others, owing to the nature of the highly-branched structure. The inner portions of the molecule may be more resistant to substitution than the outer portions.

A description of the many variables affecting hydroxyethyl starch should indicate the complexity of hydrolysis. In the remainder of this chapter, the importance of each of these variables is shown by the results of studies in man and animals.

2.2. CATABOLISM AND EXCRETION

2.2.1. PHYSICOCHEMICAL CHARACTERISTICS OF THE VARIOUS CLINICALLY-TESTED SPECIES OF HYDROXYETHYL STARCH

Throughout the remaining portions of this monograph, the individual rates of catabolism, excretion, and retention in animals and in man of a number of species of hydroxyethyl starch will be discussed. Therefore, a concise summary of their physicochemical properties will be required to give an appreciation of the interaction between the molecular weight distribution (as defined by $M_{\overline{n}}$ and $M_{\overline{w}}$) and the degree of hydroxyethyl group substitution (as defined in this instance by the MS) and their subsequent combined influence on hydrolysis *in vitro* or *in vivo.* A summary of the physicochemical characteristics of the clinically tested species of hydroxyethyl starch is presented in Table 2.1.

2.2.2. POST-TRANSFUSION SURVIVAL IN BLOOD AFTER INTRAVENOUS DOSING

2.2.2.1. THE RAPID PHASE OF CLEARANCE

When a solution of hydroxyethyl starch is injected intravenously, the blood will initially contain a heterogeneous population of polymers of various sizes.

TABLE 2.1 *Physicochemical properties of the clinically tested hydroxyethyl starches*

Species of hydroxyethyl starch	Osmolality (m0smol/l)	Inherent viscosity (dl/g)	MS*	$M_{\bar{w}}$	$M_{\bar{n}}$	Colloid osmotic pressure (cmH_2O)	$M_{\bar{w}}/M_{\bar{n}}$
HES 450/0.70†	310‡	0.18–0.30	0.70	450 000	71 000	58.5¶	6.3
HES 350/0.60	–	0.19–0.23	0.60	350 000	68 000	–	–
HES 264/0.43†	–	0.14–0.18	0.43	264 000	63 000	–	4.2
HES 200/0.60†	296‡	0.14–0.20	0.60	200 000	60 000	66.4¶	3.3
HES 150/0.70	–	0.13–0.17	0.70	150 000	41 000	–	3.7
HES 40/0.55†	285 §	0.09–0.14	0.55	40 000	31 000	77.6¶	1.3

* Hydroxyethyl groups/glucose residues contained in the parent amylopectin polymer
† Presently available
‡ Suspended in 0.9 per cent isotonic saline
§ Suspended in Ringer's lactate
¶ 6 per cent solution (w/v)

Polymers of hydroxyethyl starch whose initial size (or molecular radius) is below the renal threshold will be eliminated rapidly, causing a subsequent fall in the blood concentration (Fig. 2.4). Larger polymers of hydroxyethyl starch whose initial size does not allow immediate filtration through the glomerulus will be hydrolysed by α-amylase, and their eventual elimination from blood governed by the time required by hydrolysis to reduce the size (or molecular radius) below that of the renal threshold (see Appendix 2 for definition of the threshold limit for hydroxyethyl starch). During this later phase, α-amylase degradation is influenced by the *amount* and *pattern* of hydroxyethyl group substitution. For discussion purposes, the rapid phase of elimination from blood will be the process in which small polymers of hydroxyethyl starch are normally filtered without further hydrolysis. This first or rapid phase of clearance can be characterized by measuring: (i) the concentration of total carbohydrate in blood with the anthrone method (the concentration of hydroxyethyl starch is the difference between total carbohydrate and free glucose; see Appendix 1 for methodology); or (ii) the amount of radioactivity remaining after the injection of [^{14}C]-labelled hydroxyethyl starch. An additional technique, namely molecular exclusion filtration (MEF), may also be used. This method is able to determine the molecular weight–size distribution of polymers of hydroxyethyl starch remaining in the intravascular space after injection (see Appendix 2 for a detailed explanation of the technique). All of the above-mentioned methods, used singly or in combination, adequately describe the initial phase of hydroxyethyl starch clearance.

As clearly shown in Fig. 2.4, the concentration of hydroxyethyl starch (as determined by the anthrone method) remaining in the blood of man falls rapidly during the first 6 hours after infusion. This rapid elimination of hydroxyethyl starch from the blood-stream has also been observed to occur in dogs (Fukutome 1977; Thompson *et al.* 1962, 1967*a*, 1970, 1979) (Fig. 2.3), rabbits Irikura *et al.* 1972*k*; Kono *et al.* 1972; Tamada *et al.* 1970), and mice (Kościelak *et al.* 1977)

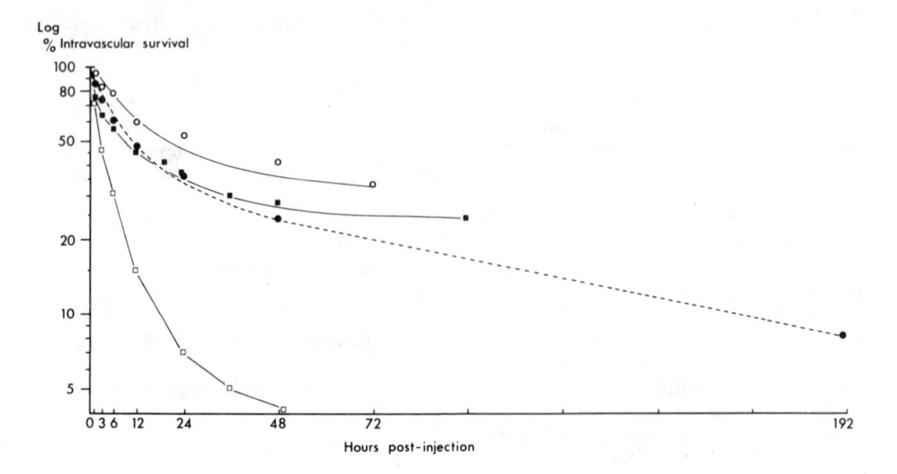

Fig. 2.4. The log % intravascular survival of HES 450/0.70 (○) (Metcalf *et al.* 1970); HES 350/0.60 (●) (Mishler 1979*c*); HES 264/0.43 (□) (Mishler *et al.* 1978*b*, 1979*b*,*c*); and HES 150/0.70 (■) (Mishler *et al.* 1978*a*) in normal healthy volunteers (see Table 2.1 for the physicochemical properties of each of the species of hydroxyethyl starch above). In each clinical study, subjects were given 54–60 g of the respective hydroxyethyl starch material.

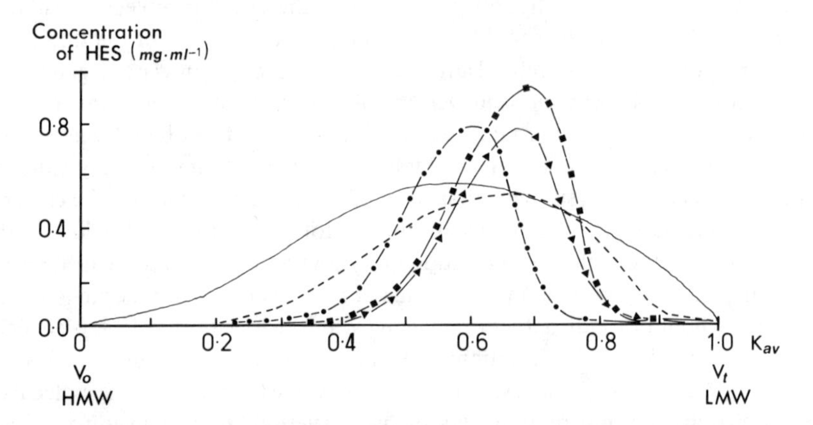

Fig. 2.5. Changes in molecular composition of HES 450/0.70 after dosing five normal men. Gel filtration on a column of Sepharose CL-4B of 50 mg HES 450/0.70 of $M_{\bar{w}}$ 450 000 (——) and a similar quantity (pooled) of HES 450/0.70 recovered from serum 2 min after injection (– – –) and one (□–□), four (△–△), and seven (●–●) weeks after dosing. In general, the distribution of molecular size present in the blood-stream is less polydispersed than was observed in the injected material. After one week there is a definite shift of molecular size towards that of a lower molecular weight (LMW) composition. However, at four weeks there are signs of a shift toward the high molecular weight (HMW) region of the injected solution, and by seven weeks this movement is quite pronounced. (From Mishler *et al.* 1980*b*).

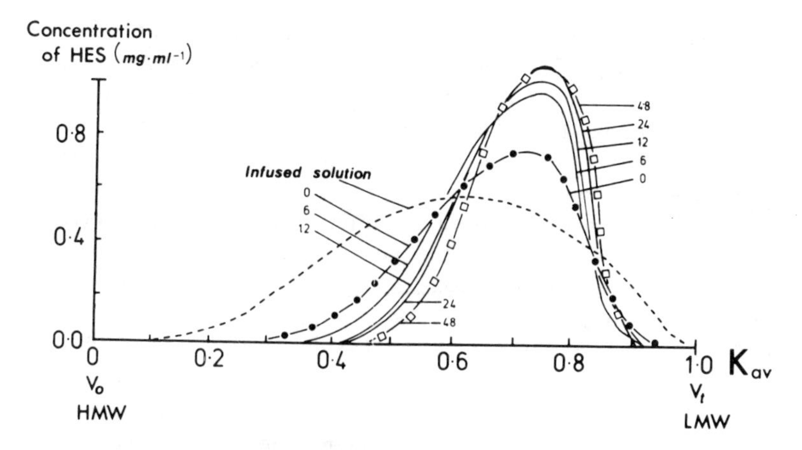

Fig. 2.6. Changes in the molecular size distribution of HES 350/0.60 after the dosing of five normal men. Gel filtration on a column of Sepharose CL-4B of 50 mg HES 350/0.60 of $M_{\bar{w}}$ 350 000 and a similar quantity (pooled) of colloid recovered from serum immediately, 6, 12, 24, and 48 hours after injection. In general, the distribution of molecular size present in blood is less polydispersed relative to the injected solution, with progressive sharpening of successive peaks. (From Mishler *et al.* 1980*c*.)

when the concentration of hydroxyethyl starch has been determined by either the anthrone method or using [^{14}C]-labelled material. The most conclusive evidence, however, to support the initial elimination of these small polymers of hydroxyethyl starch, has been provided by MEF. As depicted in Figs. 2.5 and 2.6, the molecular size distribution of polymers of hydroxyethyl starch remaining in blood immediately after the infusion clearly demonstrates the rapid removal of the small molecules filtered through the glomerulus.

2.2.2.2. A PROLONGED ELIMINATION PHASE

The next phase in the elimination of hydroxyethyl starch from the blood-stream is dependent on the hydrolysis of polymers too large to be initially filtered at the glomerulus. It is during this prolonged phase that a number of complex variables influence the rate at which polymers are catabolized. As discussed earlier, the *amount* and *pattern* of hydroxyethyl group substitution on individual glucose residues control the rate at which attack by α-amylase proceeds. Early studies in animals (Fukutome 1977; Kitamura 1972; Tamada *et al.* 1970, 1971; Yoshida *et al.* 1973) demonstrated that a direct relationship exists between the *amount* of molar hydroxyethyl group substitution in hydroxyethyl starch and the subsequent persistence in the blood-stream, i.e. the higher the amount of substitution, the longer the persistence in blood. Even though the exact *pattern* of substitution has not been determined in these studies, extrapolation of the data of Merkus *et al.* (1977) (Fig. 2.1) indicates that a greater *number* of substitutions will result in a higher frequency of multi-substituted glucose residues, enchancing unfavourable substitution combinations within the five-unit amylose substrate (Fig. 2.2), thus retarding the attack by α-amylase. During this phase, dependent on hydrolysis by α-amylase, the substitution (*amount*) of hydroxyethyl groups, rather than molecular weight, is the major determinant of persistence in blood (Tamada *et al.* 1971). This is clearly shown in man, when measuring the persistence of two species of hydroxyethyl starch, both possessing an MS of 0.70 but having significantly different molecular weights (HES 450/0.70 and HES 150/0.70) (Table 2.1 and Fig. 2.3). In this cited example, the persistence of both of these species of hydroxyethyl starch is similar, even though one species has a molecular weight three times the other.

In the discussion above, the elimination of hydroxyethyl starch has been considered after a *single* injection. In those circumstances where a species of hydroxyethyl starch is administered on multiple occasions to the same individual, the concentration of this material remaining in blood after the second injection will depend on the amount of degradation which had occurred after the initial infusion. This is depicted clearly in Fig. 2.7, where HES 450/0.70 was administered on 3 consecutive days. Twenty-four hours after each infusion, approximately 50 per cent of the initial peak concentration of HES 450/0.70 remained in the blood (Maguire *et al.* 1979*a*, 1981; Mishler *et al.* 1977*a*; Rock and Wise 1978, 1979; Strauss and Koepke 1980). The quantity of residual hydroxyethyl starch remaining in blood after the initial injection is dependent on the MS of the species of hydroxyethyl starch given, i.e. the greater the MS, the more residual material will be in blood before the next injection.

Determining the concentration of hydroxyethyl starch in blood yields little information as to the actual changes occurring in the molecular weight-size distribution. The utilization of MEF has, however, allowed interpretation of the

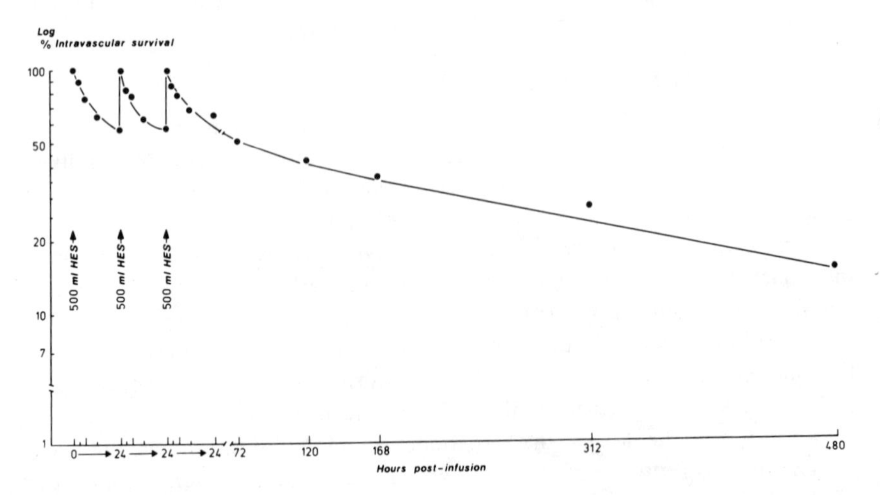

Fig. 2.7. The log % survival of HES 450/0.70 in the blood of man after three consecutive daily 30 g injections (total 90 g). (From Mishler and Beez 1979.)

effect of hydroxyethyl group substitution on catabolism mediated by α-amylase. Generally, in man, as the MS increases the longer that species of hydroxyethyl starch persists in blood (Fig. 2.4). This increased persistence in blood has been shown to correlate with a specific change in the molecular weight–size distribution after injection. Species of hydroxyethyl starch possessing an MS ⩽0.60 are rapidly catabolized after injection and the molecules remaining in blood become *smaller* relative to the infused solution (Fig. 2.6). This phenomenon has been observed for HES 40/0.55 (Irikura *et al.* 1972*k*; Oda *et al.* 1972*a*), HES 264/0.43 (Mishler *et al.* 1979*c*), and HES 350/0.60 (Mishler *et al.* 1980*c*), and is dependent on the degree (*amount*) of hydroxyethylation rather than molecular weight of the starting solution.

However, if the MS is increased to 0.70, an *intermediate* range of molecules (relative to the injected solution of hydroxyethyl starch) remain in blood (Fig. 2.5). This *intermediate* molecular weight–size distribution of polymers is observed during the phase in which the concentration of hydroxyethyl starch in blood decreases slowly (Boon *et al.* 1976; Mishler *et al.* 1980*b*) (Fig. 2.4). This finding occurs in man given HES 450/0.70 once (Mishler *et al.* 1980*b*) or on consecutive days (Misher *et al.* 1979*a*), and in rabbits administered a single injection of either HES 150/0.70 or HES 450/0.70 (Farrow *et al.* 1970; see Appendix 2). As discussed earlier, the findings of Merkus *et al.* (1977) have clearly shown that when the MS increases from 0.70 to 1.00, the numbers of di-, tri- and tetra-substituted residues of glucose increase. This increase in the proportion of multi-substituted glucose residues must then increase the likelihood that unfavourable *patterns* of substitution will take place within the five-unit amylose substrate, hindering attack by α-amylase (Fig. 2.2). Even though data on changes in *pattern* and *amount* of substitution after infusion of hydroxyethyl starch remain to be documented, species of hydroxyethyl starch possessing an MS below 0.60 are probably catabolized by α-amylase in a predictable manner. If the MS is increased beyond 0.70, however, hydrolysis by α-amylase is hindered to a greater degree, resulting in longer persistence in blood due to the production of *intermediate* rather than *smaller* polymers.

In addition to measuring the amount of hydroxyethyl starch remaining after injection as a means of comparing the effect of MS on catabolism, one may wish to determine if the appearance of free glucose correlates with the degree of hydroxyethylation. If we assume that the appearance of hyperglycaemia after injection may indicate an enhanced rate of hydrolysis then the MS also plays a major role. Hyperglycaemia is not normally observed in man when the species of hydroxyethyl starch possesses an MS equal to or greater than 0.70 (Ballinger *et al.* 1966*a*). However, species of hydroxyethyl starch having an MS below 0.60 are generally associated with the liberation of large amounts of free glucose after infusion (Adachi *et al.* 1972; Goto *et al.* 1972*b*; Irikura *et al.* 1972*k*, *q*; Mishler *et al.* 1978*a*, 1979*b*, 1980*c*; Yamasaki 1975). This inverse correlation between MS and the appearance of hyperglycaemia has also been observed in dogs (Thompson *et al.* 1960, 1962). In addition to free glucose, maltose and oligosaccharides are produced *in vitro* when hydroxyethyl starch is incubated with saliva (Farrow *et al.* 1970) or serum (Irikura *et al.* 1972*k*). Neither maltose nor oligosaccharides are detected *in vivo*, however, after infusion of hydroxyethyl starch (Irikura *et al.* 1972*k*). This may be due to the rapid excretion of these small molecules *in vivo*.

The ability to accurately predict the effect of molar hydroxyethyl group substitution (*amount* and *pattern*) on the hydrolysis of hydroxyethyl starch during this prolonged phase of elimination has been determined by measuring the concentration of hydroxyethyl starch as well as the amount of liberated free glucose and by the changes occurring in the molecular weight distribution of circulating polymers. In all these instances, the MS has been shown to influence the rate of attack by α-amylase; the higher the MS, the greater will be the persistence of hydroxyethyl starch in blood. The persistence of hydroxyethyl starch in the blood-stream can also be predicted mathematically, even though such determinations have not proved reliable. For example, up to 72 hours after a single intravenous injection has been administered to man, the disappearance of HES 450/0.70 can be described mathematically and appears to consist of three phases of elimination (rapid, intermediate, and slow half-lives) (Metcalf *et al.* 1970). Mathematical models have shown that HES 450/0.70 is eliminated from blood in a similar fashion for up to 24 hours each day after three consecutive daily infusions (Mishler *et al.* 1978*a*; Mishler and Beez 1979).

However, the *actual* long-term persistence of HES 450/0.70 in blood differs significantly from predictions made by mathematical models in which elimination was described over a relatively short period of observation (Boon *et al.* 1976; Mishler *et al.* 1980*b*). For example, the time required for the level of HES 450/0.70 to reach 1.0 per cent of the initial peak concentration was predicted mathematically to occur 2.3 and 4.5 weeks after a *single* infusion (Metcalf *et al.* 1970) or after three consecutive daily infusions (Mishler and Beez 1979), respectively. However, if the *actual* concentration of HES 450/0.70 is determined, the 1.0 per cent level is reached 17 weeks after a single infusion (Boon *et al.* 1976; Mishler *et al.* 1980*b*). The discrepancy between the disappearance of HES 450/0.70 as predicted mathematically and as observed may reside in the character of the residual material remaining in the prolonged phase of elimination. Theoretically, if the injected solution of HES 450/0.70 contains polymers in a small fractionated size range combined with a homogeneous *pattern* of hydroxyethylation, it could be assumed that disappearance from blood would be predicted adequately by an equation of the first order. The interaction between molecular weight and hydroxyethylation is highly complex and any prediction of the elimination of hydroxyethyl starch from blood can be made only exponentially at best. Mathematical models describing the elimination of HES 150/0.70 from the blood of man have also suffered from what is predicted and what is measured (Mishler *et al.* 1978*a*).

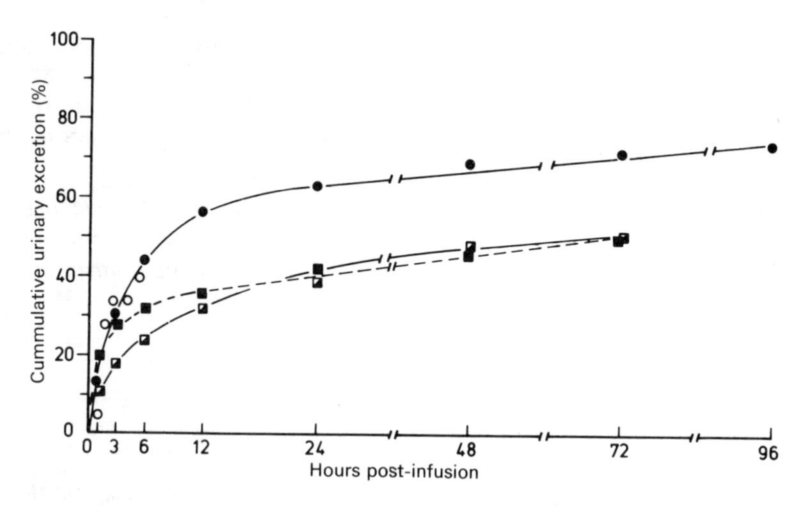

Fig. 2.8. The cumulative urinary excretion of HES 40/0.55 (○, Kitamura *et al.* 1972*a,b*); HES 150/0.70 (■, Mishler *et al.* 1978*a*); HES 264/0.43 (●, Mishler *et al.* 1978*b*, 1979*b,c*); and HES 450/0.70 (◪, Metcalf *et al.* 1970) of normal man after intravenous dosing. The total quantity of hydroxyethyl starch administered is expressed as 100%.

2.2.3. URINARY EXCRETION KINETICS AFTER INTRAVENOUS DOSING

2.2.3.1. THE RAPID CLEARANCE PHASE

As discussed earlier in Section 2.2.2.1, the first phase of elimination of hydroxyethyl starch from blood comprises the removal of small polymers whose initial size (or molecular radius) is below the renal threshold (see Appendix 2 for renal threshold limit for hydroxyethyl starch). This is shown quite convincingly by the rapid fall in the blood concentration of hydroxyethyl starch during the first 3-6 hours after injection (Figs. 2.3 and 2.4). Concomitantly, large quantities of hydroxyethyl starch appear in the urine (Fig. 2.8). In animals (Tamada *et al.* 1970) and in man (Table 2.2), the amount of the total dose of hydroxyethyl starch excreted during the first hour after injection is normally an *inverse* property of the $M_{\bar{n}}$ of the starting solution. In this context, the $M_{\bar{n}}$ can be considered to represent a population of hydroxyethyl starch polymers that would be little affected by the *amount* and *pattern* of hydroxyethylation; these polymers would generally be below the limit of the renal threshold and would be eliminated because of their low molecular weight. Confirmation of this observation has been provided by using MEF. In those species of hydroxyethyl starch that are rapidly catabolized (see Fig. 2.4), the molecular weight-size distribution of polymers initially excreted during the first hours after injection are distinctly smaller than polymers of hydroxyethyl starch voided during the slower phase of elimination (Mishler *et al.* 1980*a,d*, 1981; Oda 1972*a*). For example, polymers of HES 350/0.60 initially excreted into urine during the first hour are smaller than those voided 12–48 hours after the infusion (Fig. 2.9; see Appendix 2 for a more detailed explanation). The initial rapid excretion of polymers of hydroxyethyl starch is due to a molecular weight (or radius) that falls below the renal threshold and appears little affected by the MS of the starting solution.

TABLE 2.2 *The inverse relationship between the $M_{\bar{n}}$ and the quantity of hydroxyethyl starch voided into the urine 1 hour after an intravenous injection. Results are expressed as a percentage of the total dose given.**

Species of hydroxyethyl starch	$M_{\bar{n}}$	Percentage total dose excreted	Reference
HES 40/0.55	31 000	14.5	Kitamura *et al.* (1972*a,b*)
HES 264/0.43	63 000	13.5	Mishler *et al.* (1979*b*)
HES 350/0.60	68 000†	12.7	Mishler *et al.* (1981)
HES 450/0.70	71 000	8.6–9.9	Mishler *et al.* (1977*b*)
			Metcalf *et al.* (1970)

* The subject population consisted of normal healthy volunteers or patients in hospital for minor operations
† Approximate value

2.2.3.2. A PROLONGED PHASE OF ELIMINATION

In Section 2.2.2.2, it was shown that the elimination of hydroxyethyl starch from blood after the rapid phase of clearance was dependent on the hydrolysis of polymers too large to be initially filtered by the glomerulus. It is during this later or second phase of elimination that the *amount* and *pattern* of hydroxyethyl group substitution controls the rate of attack by α-amylase. This can be seen quite clearly when comparing the clearance of various species of hydroxyethyl starch from the blood (Fig. 2.4) with the subsequent cumulative urinary excretion (Fig. 2.8). Species of hydroxyethyl starch (HES 450/0.70 and HES 150/0.70) possessing the highest MS persist in blood the longest and consequently have the lowest overall rate of cumulative urinary excretion (Mishler *et al.* 1977*b*). This relation appears to be independent of the $M_{\bar{w}}$ of the infused solution of hydroxyethyl starch. The rate of urinary excretion during this prolonged phase of elimination is thus dependent on the ability of α-amylase to catabolize large polymers of hydroxyethyl starch to a size (or molecular radius) which falls below the renal threshold. Hydrolysis, in turn, is influenced by the *amount* and *pattern* of hydroxyethyl group substitution. Polymers of hydroxyethyl starch excreted into urine during this second phase of elimination thus have a relatively uniform size (Mishler *et al.* 1981; Oda 1972*a*), a size distinctly larger than those small polymers excreted during the rapid phase of clearance (Fig. 2.9). In those species of hydroxyethyl starch that are catabolized rapidly (MS $\leqslant$ 0.60), the polymers remaining in blood become *smaller*, and with time approach a molecular size distribution similar to those molecules that are excreted into the urine (Fig. 2.9).

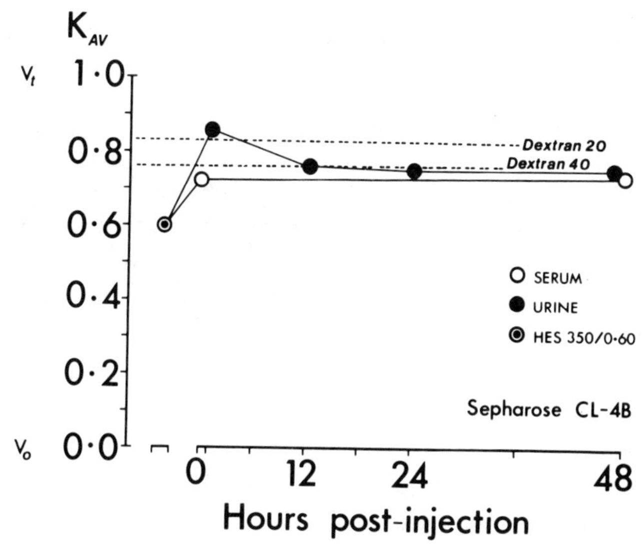

Fig. 2.9. Analysis of the changes occurring in the molecular size distribution of polymer fragments of HES 350/0.60 recovered at various times after injection either from the serum or urine of normovolaemic subjects (Mishler 1980*c*). For comparison, standard preparations of dextran 20 (K_{av} 0.83, $r = 3.2$ nm) and dextran 40 (K_{av} 0.76, $r = 4.5$ nm) were separated on the identical column of Sepharose CL-4B (see Appendix 2 for details).

This is not true of species of hydroxyethyl starch that are catabolized slowly (MS $\geqslant$ 0.70). As shown in Fig. 2.5, an *intermediate* rather than *smaller* molecular weight-size distribution is seen in polymers remaining in blood after four weeks. The occurrence of this *intermediate* molecular weight-size distribution in blood must ultimately decrease the rate at which polymers of hydroxyethyl starch are excreted into the urine. Thus, during the prolonged phase of elimination, large polymers are excreted into urine as soon as their size falls below the renal threshold. The attainment of this size is dependent on the rate at which α-amylase is able to catabolize these large polymers, which in turn is influenced by the *amount* and *pattern* of hydroxyethyl group substitution.

2.3. TISSUE STORAGE OF HYDROXYETHYL STARCH

The fate of hydroxyethyl starch has been studied extensively in animals by means of cytochemical staining and by using [^{14}C]-labelled material. This latter technique will be discussed in detail in the next section. In man, however, cytochemical staining of tissue sections appears to be the only method available to study retention, even although safer radio-labelled material may become available.

2.3.1. THE FATE OF [^{14}C]-LABELLED HYDROXYETHYL STARCH

The chemical synthesis of hydroxyethyl starch includes the attachment of a hydroxyethyl group (contributed by ethylene oxide) through an *ether* linkage to the hydroxyl group on either C2, C3, or C6 of the anhydroglucose residue (see Section 1.2). This reaction sequence not only allows the substitution of radio-active ethylene oxide (1,2-[^{14}C]-ethylene oxide) as a means of thoroughly documenting the fate of subsequently labelled [^{14}C]-hydroxyethyl starch in the body, but also yields specific information on the stability of *ether* linkages *in vivo.* Cleavage of this *ether* linkage with subsequent release of ethylene glycol should contribute significant quantities of $^{14}CO_2$ to the expired air. As reported by Gessner *et al.* (1960), ethylene glycol produces CO_2 as the major metabolite in rats and rabbits. The recognized chemical stability of *ether* linkages contrasts with the ready hydrolysis of starch α-1,4 glycosidic linkages by α-amylase.

2.3.1.1. INTRAVENOUS ADMINISTRATION

2.3.1.1.1. Mice

Six hours after the injection of [^{14}C]-labelled HES 450/0.70, 30-51 per cent of the total administered dose remained in the blood-stream while 26-33 per cent was excreted in the urine (Bogan *et al.* 1969; Suzuki *et al.* 1971). During this same period of observation, 0-2, 3-9 and 8-9 per cent of the amount injected was detected in faeces, liver, and muscle, respectively. Less than 4 per cent of the amount injected appeared in kidney, intestine, lungs, heart, spleen, testis, or brain (Suzuki *et al.* 1971). The mean total of $^{14}CO_2$ collected in expired air after 6 hours was 0.05 per cent of the quantity injected, and approximately 60-70 per cent of this was collected within the first 2 hours after injection (Bogan *et al.* 1969). The amount of $^{14}CO_2$ expired by the sixth hour was undetectable. Contrastingly, mice given D-glucose-1-[^{14}C] intravenously, exhaled 54-60 per cent of the total administered dose during a 6-hour collection period (Bogan *et al.* 1969). In mice dosed with [^{14}C]-labelled HES 40/0.55, approximately 40 per cent of the total amount injected was excreted in the urine 6 hours after injection (Takai 1972). No activity was detected in the faeces.

Twenty-four hours after the injection of [^{14}C]-labelled HES 450/0.70, 17 per cent of the total amount given remained in the intravascular space of mice, 37 per cent was excreted in the urine, while 3, 8, and 13 per cent were detected in liver, faeces, and muscle, respectively (Suzuki *et al.* 1971). Less than 1 per cent of the amount injected remained in kidney, intestine, lung, spleen, heart, testis, or brain. Twenty-four hours after the dosing of mice with [^{14}C]-labelled HES 40/0.55, 19 and 74 per cent of the amount injected was excreted in faeces and urine, respectively (Takai 1972). Twenty-four hours after the injection of either [^{14}C]-labelled HES[1] 200/0.55, HES[1] 200/0.65 or HES[1] 200/0.80, most of the total dose given had been eliminated from blood, spleen, liver, kidney, heart, and lungs (Kościelak *et al.* 1977). Approximately 13-17 per cent of the amount injected was detected in muscle.

[1] This species of hydroxyethyl starch was synthesized by the investigators, not by a commerical manufacturer.

Ninety-six hours after the dosing of mice with [^{14}C]-labelled HES 450/0.70, the [^{14}C]-activity in testis, spleen, blood, and muscle was 0.4, 0.5, 2, and 10 per cent of the amount injected (Suzuki *et al.* 1971).

Sixty days after the dosing of mice with either [^{14}C]-labelled HES[1] 200/0.55 or HES[1] 200/0.80, less than 0.5 per cent of the amount injected was detected in blood, spleen, liver, kidney, heart, or lung (Kościelak *et al.* 1977). Small, detectable amounts of [^{14}C]-activity were still present in muscle in direct relation to the MS of the injected material.

2.3.1.1.2. Rats

Six hours after the injection of [^{14}C]-labelled HES 450/0.70, 39–49 per cent of the amount administered appeared in blood while 18–30 per cent was excreted in the urine (Suzuki *et al.* 1971). During this same period of observation, 2–3 and 3–7 per cent of the amount injected was detected in liver and muscle, respectively. Less than 2 per cent of the total dose remained in faeces, intestine, kidney, lung, spleen, heart, testis, or brain. In rats dosed with [^{14}C]-labelled HES 450/0.70, less than 0.1 per cent of the amount injected was expired as $^{14}CO_2$ during the first 6 hours of collection (Thompson *et al.* 1970, 1979). In rats given [^{14}C]-labelled hydroxyethyl starch, less than 0.5 per cent of the total radioactivity was detected in bile, 6 hours after injection (Ryan *et al.* 1972).

Twenty-four hours after the injection of [^{14}C]-labelled HES 450/0.70, 17–21 per cent of the amount administered remained in the intravascular space, 38–44 per cent was excreted in urine, while 1–4, 3, and 2–17 per cent was detected in muscle, liver, and faeces, respectively (Suzuki *et al.* 1971; Thompson *et al.* 1970, 1979).

In rats dosed with [^{14}C]-labelled HES 450/0.70, urinary excretion averaged 41 per cent at 48 hours and 43 per cent at 72 hours (Thompson *et al.* 1970, 1979), while the [^{14}C]-activity in blood fell progressively from 6 per cent at 48 hours to 1.6 per cent at 96 hours (Suzuki *et al.* 1971). The mean cumulative faecal excretion was 53 per cent in three weeks (Thompson *et al.* 1970, 1979).

Twenty-eight days after injection of [^{14}C]-labelled HES 450/0.70, 74.1 and 10.9 per cent of the total administered dose was detected in urine and faeces, respectively, while the carcass retained 12.5 per cent with 1.7 and 1.9 per cent remaining in spleen and liver, respectively (Hulse *et al.* 1980).

Thompson *et al.* (1970, 1979) have calculated the tissue persistence of [^{14}C]-labelled HES 450/0.70 in various organs after intravenous dosing. The half-life of radioactivity in plasma, kidney, carcass, spleen, heart, and liver is about 7.2, 18.9, 50, 64, 95.2, and 132 days, respectively. In these animals, the spleen [^{14}C]-concentration rose to a peak at one week of thirty-five times the initial average whole-body specific activity, but fell progressively so that after ten weeks the spleen contained only 0.37 per cent of the injected dose and weighed 0.2 per cent of the body weight. The [^{14}C]-activity of liver increased to a peak

[1] This species of hydroxyethyl starch was synthesized by the investigators, not by a commercial manufacturer.

at one week of 2.5 times the initial average whole-body specific activity, but steadily fell to 2 per cent after ten weeks, concomitant with the liver weighing only 3.3 per cent of the body weight. The [^{14}C]-activity in the remainder of the carcass decreased progressively from 16 per cent of the injected dose at six weeks to 5 per cent at ten weeks (Thompson *et al.* 1970, 1979).

2.3.1.1.3. Dogs

Twelve hours after dosing dogs with [^{14}C]-labelled HES 450/0.70, 15 per cent of the initial level remained in the blood-stream, while 82 per cent had been excreted (cumulative value) in the urine (Bogan *et al.* 1969). Ninety-six hours after injection, 2.3 per cent of the initial level was detected in blood, while after 22 days, only 1.0 per cent was retained intravascularly.

2.3.1.1.4. Rabbits

Twenty-four hours after the injection of [^{14}C]-labelled HES 450/0.70, 33 per cent of the amount administered remained in the blood-stream, while 44 per cent was excreted in urine (Suzuki *et al.* 1971). Ninety-six hours after injection, 14 per cent of the [^{14}C]-activity was detected in blood, while 54 per cent had been excreted in urine. Twenty-two days after injection, 1.0 per cent of the amount injected remained in the intravascular space. (See Table 2.3 for comparison between various animal species in regard to [^{14}C]-hydroxyethyl starch activity in blood, urine, and faeces 24 hours after intravenous dosing.)

TABLE 2.3 *The distribution of [^{14}C]-labelled hydroxyethyl starch in blood, urine, and faeces as a percentage of the total administered dose in mice, rats, dogs, and rabbits 24 hours after a single intravenous injection*

Species	Species of [^{14}C]-labelled hydroxyethyl starch	Distribution (% of injected dose)			
		Blood	Urine	Faeces	IT_{50} (hours)
Mice	HES 450/0.70	17	37	8	≈ 6
	HES 40/0.55	–	74	19	–
Rats	HES 450/0.70	17–21	38–44	2–17	≈ 3–6
Dogs	HES 450/0.70	9	82*	–	≈ 1
Rabbits	HES 450/0.70	33	44	–	≈ 8

* Cumulative urinary excretion after 12 hours

2.3.1.2. ORAL ADMINISTRATION

In rats dosed orally with [^{14}C]-labelled hydroxyethyl starch[1] virtually all the [^{14}C]-activity (> 97 per cent) was excreted after 96 hours (Ryan *et al.* 1972). In the first 24 hours after dosing, 7 and 36 per cent of the amount consumed was excreted in urine and faeces, respectively. During this same period of observation, only 11 per cent was expired as $^{14}CO_2$. Ninety-six hours after feeding, the cumulative excretion via urine and faeces was 14.3 and 61.2 per cent, respectively.

[1] Synthesized by the investigators.

The cumulative amount of $^{14}CO_2$ expired during this period was 21.9 per cent of the dose administered. In contrast to these findings, 24 hours after dosing rats orally with [^{14}C]-labelled *unsubstituted* starch, 7.5 and 15.5 per cent of the amount given was detected in urine and faeces respectively, while 45.3 per cent appeared as expired $^{14}CO_2$ (Ryan *et al.* 1972). Anderson *et al.* (1963) have also reported large quantities of expired $^{14}CO_2$ after oral dosing of rats with [^{14}C]-*unsubstituted* starch.

2.3.1.3. INTRAPERITONEAL ADMINISTRATION

2.3.1.3.1. Mice

In two mice dosed with [^{14}C]-labelled HES 450/0.70, 0.04 and 0.05 per cent of the amount injected was exhaled as $^{14}CO_2$ 6 hours after collection (Bogan *et al.* 1969).

2.3.1.3.2. Rats

Six hours after the dosing of two rats with [^{14}C]-labelled HES 450/0.70, 0.03 and 0.05 per cent of the amount injected was expired as $^{14}CO_2$ (Bogan *et al.* 1969). In rats given [^{14}C]-labelled hydroxyethyl starch,[1] 95 per cent of the radioactivity appeared in the urine by 24 hours after injection with a further 1-2 per cent in the next 72 hours (Ryan *et al.* 1972). After 24 hours, 2 per cent of the [^{14}C]-activity was detected in the faeces.

2.3.2. HISTOLOGICAL ALTERATIONS

Extravasated polymer, lost from the intravascular space but not appearing in voided urine, accumulates to the same extent with hydroxyethyl starch and dextran (Thompson *et al.* 1970, 1979). For example, 24 hours after injection, 30 per cent of HES 450/0.70 and 37 per cent of dextran 70 were extravascular in dogs. Twenty-four hours after injection, 43 per cent of hydroxyethyl starch and dextran were extravascular in rats. In man, approximately 25-30 per cent of the total administered dose of either HES 264/0.43 (Mishler *et al.* 1979*b*) or HES 450/0.70 (Metcalf *et al.* 1970) remains unaccountable (total dose − hydroxyethyl starch in blood and urine = unaccountable fraction) and is assumed to be extravascular. This extravasated fraction appears constant throughout the sampling period (72-96 hours after injection) and is similar to animal studies reported above. In Section 2.3.1, the clearance of extravasated [^{14}C]-labelled hydroxyethyl starch from major viscera was documented. This section will focus on the histological changes produced in major viscera by the presence of extravasated hydroxyethyl starch.

In most studies cited in this section, histopathological changes taking place after the infusion of hydroxyethyl starch were depicted by light and electron microscopy. Cytochemical techniques were also used to document the presence

[1] Synthesized by the investigators.

of intracellular water-soluble polysaccharide granules. For example, the use of alcoholic periodate-oxidation of starch and dextran followed by aqueous Schiff's reagent (periodic-acid Schiff [PAS]), clearly stains intracellular hydroxyethyl starch or dextran granules red or red-purple (Thompson 1963). In distinguishing hydroxyethyl starch and dextran (both water-soluble) from glycogen (water-insoluble) in tissue sections, similar areas in the slides treated with alcoholic periodate, and therefore containing hydroxyethyl starch or dextran and glycogen, should be compared with slides treated with water before periodate-oxidation in which hydroxyethyl starch or dextran are washed out (Thompson 1963).

2.3.2.1. AFTER ACUTE ADMINISTRATION

2.3.2.1.1. Mice, rabbits, dogs, and rats

Histological changes were examined in mice, rabbits, and dogs given 1.8 g/kg of either HES 40/0.55 or dextran 70 (Irikura *et al.* 1972*k*, *p*). Stained materials in hepatic cells as well as tubular epithelium was less in animals treated with HES 40/0.55 than in animals given dextran 70. Vacuolization, hepatic-cell swelling and hydropic swelling of tubular epithelium were also less pronounced in animals receiving HES 40/0.55.

Morphological changes were examined in dehydrated rabbits given 3.6 g/kg of either HES 40/0.55 or dextran 40 (Yamasaki 1975). Kidney sections taken 3 hours after transfusion revealed multiple small cytoplasmic vacuoles under the brush border of proximal tubular epithelium. These vacuoles increased in size and number and coalesced to form large vacuoles, some of which contained an electron-dense material by PAS staining in samples of renal tissue obtained 12 hours after infusion. Morphological alterations were always more pronounced in tissue specimens taken from dextran 40-treated animals.

In dogs given 1.2 g/kg of HES 450/0.70, characteristic PAS-positive starch granules were found in the renal intravascular and interstitial spaces 24 hours after the infusion (Murphy *et al.* 1965). Granules were also found within the proximal tubular cell cytoplasm and within the tubular lumen. However, there was no evidence of tubular vacuolisation.

The kidneys of rats, exchange-transfused down to a haematocrit of 20 per cent with either dextran 60 or HES 450/0.70, showed no pathological changes when examined by light microscopy (Hölscher *et al.* 1975).

2.3.2.1.2. Monkeys

Sections of kidney, lung, lymph nodes, spleen, and liver were examined in monkeys before and after receiving an HES 150/0.70 previously frozen blood admixture, and in monkeys not given HES 150/0.70 and serving as histopathological controls (Weatherbee *et al.* 1974*c*). Formalin- and alcohol-fixed tissues were stained with haematoxylin and eosin and with PAS (with or without diastase digestion). The results showed no consistent differences between control and test animals. PAS-positive material was noted in macrophages both in control and in test animals, and these authors concluded that there was no evidence that hydroxyethyl starch of low molecular weight persists in tissues after acute administration.

2.3.2.1.3. Man

Liver sections obtained by biopsy were examined by light and electron microscopy 30 min to 28 days after injection in patients given 60 g HES 450/0.70 while undergoing abdominal surgery (Jesch *et al.* 1979). Single intracellular vacuoles were shown by electron microscopy 30 min after injection only in Kupffer cells. However, 6–28 days after injection, intracellular vacuoles were demonstrated in parenchymal liver cells, Kupffer cells, interstitial histiocytes, and to a lesser degree in the cells of the small bile ducts. The methods used in this study (e.g. lack of a control group) and their interpretation have been recently criticized (Thompson 1979).

2.3.2.2. AFTER CHRONIC ADMINISTRATION

2.3.2.2.1. Rabbits

Liver weights were not increased in rabbits given 1.8 g HES 40/0.55 for four consecutive days (Irikura *et al.* 1972*q*). The wet or relative weights of heart and liver in animals treated for one week (3 g HES 140/0.65/kg per day), and heart, liver, kidneys, adrenals, and spleens of animals treated for either three weeks (3 g/kg per day of either HES 140/0.65, HES 450/0.70, or dextran 60) or thirty days (3.6 g/kg per day of either HES 40/0.55 or dextran 70) were significantly higher when compared with controls (Hölscher 1972, 1975; Lindblad 1970; Irikura *et al.* 1972*g*). No significant change in organ weight, with the exception of adrenals, however, was noted in animals treated with 1.8 g HES 40/0.55/kg per day for thirty days (Irikura *et al.* 1972*g*). Some organ weights were increased in rabbits treated intravenously with 0.7 or 2.1 g HES 450/0.55/kg per day for thirty days (Irikura *et al.* 1972*l*).

Organ weights of heart, liver, and spleen were increased while the weight of the pancreas decreased in rabbits administered either 0.9, 1.8, or 3.6 g/kg per day HES 40/0.55 for three months (Irikura 1972).

In animals administered 6 g/kg per day of either HES 40/0.55 or dextran 70 intravenously for four consecutive days, persistence of PAS-stained material in splenic red pulp was longer in the group receiving HES 40/0.55 (Irikura *et al.* 1972*p*). No histopathological changes were noted in the spleens of either group, and long-term persistence was not observed in the red pulp when examined by light and electron microscopy. In animals treated with dextran 70, however, there were moderate to massive amounts of PAS-stained granules both in tubular epithelium and glomeruli. Hydropic swelling of tubular epithelium was severe in the dextran 70-treated rabbits. Hepatocytes of animals treated with dextran 70 displayed abnormal morphology when examined by electron microscopy, although slight changes were noted with light microscopy.

Large amounts of fluid were found in the serous cavities of rabbits treated intravenously for one week (Lindblad 1970), ten days (Lindblad and Falk 1976), three weeks (Hölscher 1972, 1975; Lindblad 1970), or thirty days (Irikura *et al.* 1972*g*) with 1.8-3.6 g/kg per day of either HES 40/0.55, HES 140/0.65, HES 450/0.70, dextran 60, or dextran 70.

Light microscopy revealed that animals given either 3 g/kg per day of either HES 140/0.65, HES 450/0.70, or dextran 60 for three weeks, or 3.6 g/kg per day of either HES 40/0.55 or dextran 70 for thirty days, had severe focal hydropic alterations of the epithelium of the proximal convoluted tubules with evidence of necrosis and desquamation of necrotic cells and swelling of glomerular tufts (Hölscher 1972, 1975; Irikura *et al.* 1972*g*; Lindblad 1970). Slight hydropic swelling of tubular epithelium was observed in animals given 1.8 g HES 40/0.55/kg per day for thirty days (Irikura *et al.* 1972*g*). Hyaline casts and erythrocytes were found in the Bowmann capsular spaces and the tubular lumina in animals treated with 3 g/kg per day of either HES 450/0.70 or dextran 60 for three weeks (Hölscher 1972, 1975; Lindblad 1970). Foam-cell production in the whole liver was pronounced and numerous necrotic parenchymal cells and extravasated erythrocytes were visible. The lungs of some rabbits given 3 g/kg per day of either HES 450/0.70 or dextran 60 for three weeks showed hyaline deposits in the lumina of the alveoli, vacuolization of the alveolar epithelia, and focal infiltrations and erythrocytes and round cells (Hölscher 1972, 1975). The spleens in these animals displayed dilated sinusoids. The lymph follicles were atrophic concomitant with endothelial degeneration (Lindblad 1970).

Electron microscopy revealed numerous vacuoles located under the brush border and in the mid-zonal area of the kidney, and less commonly in the basal portion of the proximal tubular cells in animals given 3 g/kg per day of either HES 450/0.70 or dextran 60 (Hölscher 1972, 1975). These cells contained fine granules or filaments of electron-dense material and dark, homogeneous, irregularly shaped masses. The marked vacuolisation of the epithelial cells of the proximal tubules is due to an increased reabsorption of large quantities of excreted hydroxyethyl starch or dextran 60 (Hölscher 1972). Numerous vacuoles of varying size containing granular and filamentous dense material were observed in the cells of the liver and endothelial cells of the sinusoids of animals given 3 g/kg per day of either HES 140/0.65, HES 450/0.70 or dextran 60 for three weeks (Hölscher 1972, 1975; Lindblad 1970) and in animals treated with 3.6 g/kg per day of either HES 40/0.55 or dextran 70 for thirty days (Irikura *et al.* 1972*g*). Vacuolation of liver cells was not observed, however, when animals were given 1.8 g/kg per day HES 40/0.55 for thirty days (Irikura *et al.* 1972*g*).

In rabbits given 3 g/kg per day of either HES 450/0.70 or dextran 60 for ten consecutive days, Lindblad and Falk (1976) concluded that HES 450/0.70 is not completely eliminated within one year after dosing, and that hydroxyethyl starch gives rise to longer-lasting morphological changes than does dextran. Additional studies (Irikura *et al.* 1972*p*; Thompson 1977*a*,*b*) comparing cellular alterations produced by massive infusions of either hydroxyethyl starch or dextran have not however, supported the above conclusions. For example, rabbits dosed with 3-3.6 g/kg per day of HES 40/0.55, HES 450/0.70, dextran

60, or dextran 70 for thirty days to three weeks had similar degrees of histological alterations (Hölscher 1972, 1975; Irikura *et al.* 1972*g*). There was no evidence presented indicating that hydroxyethyl starch gave rise to more severe or more longer-lasting histological changes.

Hydropic swelling of renal tubular epithelium and hepatocytes, vacuolisation of splenic red pulp, swelling of reticular cells of lymph nodes as well as epithelial cells of the epididymis, and a decrease in the number of spermatozoa were noted in rabbits given 0.9, 1.8, or 3.6 g/kg per day of HES 40/0.55 for three months (Irikura 1972).

2.3.2.2.2. Rats and mice

Morphological changes in reticulum cells of the lymph nodes were examined by cytochemical techniques, as well as by light and electron microscopy, in rats given either 0.8 g HES 40/0.55/kg per day or 0.5 g dextran 40/kg per day for four consecutive days (Paulini and Sonntag 1976). Both polymers were shown to be stored in cells of the reticuloendothelial system (RES). Reticulum-cell necrosis was observed after storage of dextran 40; however, no such findings occurred in animals treated with HES 40/0.55. The authors concluded that both substances did not block the RES. An additional study (Irikura *et al.* 1972*m*) has also substantiated this last observation. In mice treated intravenously with 6 g/kg per day of HES 40/0.55 for four days, the carbon clearance test, as well as phagocytic activity of leucocytes were not significantly different when compared with control animals receiving lactated Ringer's solution.

In male or female mice treated intravenously with either 3 g or 5 g/kg per day of HES 140/0.65 for three weeks, the weight gain was significantly lower in the male group receiving 5 g/kg per day than in the control group (Lindblad 1970). There were no differences between the other groups. The liver weights in all treated groups were higher than in control animals. All livers in male or female mice receiving 3 g/kg per day displayed severe alterations (dilated sinusoids with cellular proliferations and infiltrations.) Liver cells were atrophic and most of them contained very little or no glycogen. Degenerative alterations and necrosis were also noted. The spleens showed dilated sinusoids with very few extravasated erythrocytes. Hydropic degeneration of tubular epithelium was noted only in the kidneys of two animals. In the animals receiving 5 g/kg per day, the changes in liver, spleen, and kidney were similar but more marked than in animals given 3 g/kg per day. Electron microscopy revealed pronounced vacuolisation of epithelial cells of the glomeruli.

In rats given either 60, 120, or 240 mg/kg per day of HES 450/0.70 for twenty-eight days, slight numbers of PAS-positive granules were observed in Kupffer cells two months after the end of the infusion period in animals treated with more than 120 mg/kg per day (Odaka *et al.* 1971*a*).

2.3.2.2.3. Dogs

Liver enlargement was noted on gross examination in male and female animals receiving 3 g HES 140/0.65/kg per day for three weeks (Lindblad 1970). All treated dogs showed a marked, widespread, hydropic degeneration of liver cells. There was, however, no difference between those animals killed immediately after the end of the infusion period, and those killed four and a half months later. Glomerular capillary loops were dilated in treated dogs, concomitant with proliferation of endothelial cells and sometimes with partial destruction of glomeruli. Very few proximal tubules displayed hydropic degeneration. Reticular cells of the splenic pulp were swollen and vacuolated and nuclei had often degenerated. Lymph-node sinuses were dilated and large numbers of swollen and degenerated phagocytes were present. The endothelial adrenal cells of zona fasciculata and zona reticularis were swollen and degenerated. In female animals, the cytoplasm of stromal cells in the ovary were focally swollen and vacuolated and nuclei were pyknotic. All other organs examined were histologically normal.

The kidneys, hearts, and lungs of dogs killed immediately after receiving either 7.8-15 g/kg HES 450/0.70 or 8.7-15 g/kg dextran 70 over 2-5 days weighed significantly more than the organs of control animals, or dogs killed three or more days after the last injection of either HES 450/0.70 or dextran 70 (Thompson 1963; Thompson *et al.* 1970, 1979). Liver weights were similar in all dogs, but spleen weights were significantly greater in animals killed several days after the administration of HES 450/0.70 than after dextran 70. Water-soluble PAS-positive hydroxyethyl starch and dextran granules were noted in intravascular and interstitial spaces, parenchymal liver cells, proximal renal tubular cells, and phagocytes in liver, spleen, lymph nodes, and other organs of dogs killed immediately after the last injection of either polymer. In these animals, the proximal renal tubular epithelial cells were swollen and filled with granulovacuoles. Six days after the last injection of either 14.4-27 g HES 450/0.70 or 14.4 g dextran 70/kg over 12-18 days, swelling of the proximal tubular cells and PAS-positive granules were observed only in areas of interstitial nephritis. Small amounts of PAS-positive granules were found in Kupffer cells, splenic and lymph node phagocytes, and tissue macrophages to the same degree in animals injected with HES 450/0.70 or dextran 70 six days before death. Contrary to the findings of Lindblad (1970), dogs killed 30 or 60 days after the last injection of either HES 450/0.70 or dextran 70 had no PAS-positive granules in any tissue examined. This last observation requires further elaboration. Animals in the Lindblad study were given a total of 21 g/kg HES 140/0.65 over twenty-one days, while animals in the study conducted by Thompson and his colleagues received a total of 14.4-27 g HES 450/0.70/kg over twelve to seventeen days.

2.4. SUMMARY

The presence of hydroxyethyl starch in the blood-stream of animals and of man triggers a complex series of events. Firstly, low molecular weight polymers are eliminated rapidly because their size or molecular radius falls below the threshold of glomerular filtration (Mishler 1980*d*). Secondly, larger polymers are reduced in size by the action of α-amylase; the rate of attack is inversely related to the number of hydroxyethyl-group substitutions contained within the parent amylopectin molecule (Mishler 1980*d*). Once the molecular weight or radius is reduced to allow for filtration at the glomerulus, these molecules will leave the intravascular space. Finally, a portion of the injected dose will become extravasated and retained in the RES (Lindblad 1970; Thompson 1963; Thompson *et al.* 1970, 1979).

The eventual breakdown and elimination of the tissue-stored hydroxyethyl starch may include the action of cytoplasmic lysosomes (Hölscher 1972). This is shown by two lines of evidence. Firstly, hydroxyethyl starch stored in liver is hydrolysed in a similar manner to carbohydrates and glycogen (Irikura *et al.* 1972*k*). Secondly, PAS-positive granules are absent weeks after the infusion of large quantities of hydroxyethyl starch (Thompson *et al.* 1970, 1979). It should be emphasized here that storage of polymers in tissue and organ dysfunction should be distinguished from each other. Even under conditions where massive amounts of hydroxyethyl starch are administered and tissues appear abnormal, normal organ function remains (see Section 3.9).

3. Effects produced by the acute and chronic administration of hydroxyethyl starch

3.1. INTRODUCTION

The persistence of hydroxyethyl starch in blood appears to be a direct consequence of the *amount* and *pattern* of hydroxyethylation (Section 2.2). Even though the parent amylopectin polymer resembles the highly-branched structure of native glycogen, prolonged persistence of hydroxyethyl starch in blood or storage in tissues could potentially result in toxic manifestations. In addition, little is known of the mechanism(s) by which hydroxyethyl starch is removed (or degraded) from tissues once it has become intracellular, although lysosomes are believed to participate in this process (Chapter 2, p. 53).

In this chapter I shall specifically consider whether high doses of hydroxyethyl starch, given acutely or chronically, would be injurious to the host.

3.2. ACUTE AND CHRONIC TOXICITY

3.2.1. ACUTE TOXICITY

The lethal dose at which 50 per cent of the injected animals survive (LD_{50}) is over 4.5 g/kg when HES 450/0.70 is given intravenously to mice, rabbits, or cats (Table 3.1). This would represent a total dose of over 315 g if given to a 70 kg man. If a smaller and more rapidly catabolized species of hydroxyethyl starch (HES 40/0.55) is administered directly into the blood of mice, rats, or rabbits, the LD_{50} is over 15.7 g/kg. This latter dose represents the injection of over 1099 g into a 70 kg man. If HES 40/0.55 is given intraperitoneally or subcutaneously to mice or rats, the LD_{50} is over 18.0 g/kg. Mice appear to tolerate greater doses of HES 40/0.55 when given orally than do rats. Survival after acute haemodilution will be dealt with in Section 4.2.

3.2.2. CHRONIC TOXICITY TESTING

In rabbits given HES 40/0.55 intravenously, all animals survived a dose of either 1.8 g/kg per day for four days or 0.9 g/kg per day for three months (Table 3.2). All rats survived a dose of either 1.5 or 3.0 g HES 40/0.55/kg per day for four days. Seventy-eight to 89 per cent of rabbits survived a dose of of either 1.8 or 3.6 g HES 40/0.55/kg per day when administered over thirty days to three months.

TABLE 3.1 *Acute toxicity testing of hydroxyethyl starch in animals*

Test animals	Species of hydroxyethyl starch	Dose (LD_{50}, g/kg)	Route	Total equivalent dose for a 70kg man (g)	Reference
Mice	HES 450/0.70	>6.0	i.v.	>420	Roberts and Pagones (1965)
Rabbits	HES 450/0.70	>5.4	i.v.	>378	
Cats	HES 450/0.70	>4.5	i.v.	>315	
Mice	HES 40/0.55	>12.0	p.o.		Irikura *et al.* (1972*n*)
		>18.0	s.c.		
		>18.0	i.p.		
		>21.6	i.v.	>1512	
Rats	HES 40/0.55	>3.0	p.o.		
		>18.0	s.c.		
		>18.0	i.p.		
		>15.7	i.v.	>1099	
Rabbits	HES 40/0.55	24.0	i.v.	1680	

All rabbits survived a dose of 0.1 g/kg per day HES 81/0.60 for three months. When HES 140/0.65 was administered over three weeks all mice and all dogs survived a dose of either 3 or 5 g/kg per day or 3 g/kg per day, respectively. When HES 140/0.65 was given over 7–21 days, 95 and 50 per cent of rabbits survived a dose of 3 g/kg per day, respectively.

When HES 450/0.55 was suspended in a solution of either 0.9 per cent sodium chloride or Ringer's lactate, 83–100 per cent of rabbits survived a dose of 0.7 or 2.1 g/kg per day for 30 days. When HES 450/0.55 was suspended in Ringer's lactate, 17 per cent of rabbits dosed with 6.3 g/kg per day for thirty days survived, while none survived the same dose when HES 450/0.55 was suspended in a solution of 0.9 per cent sodium chloride.

Seventeen per cent of rabbits survived a dose of 3 g/kg per day of HES 450/0.70 for three weeks.

All animals chronically dosed with any species of hydroxyethyl starch will retain this colloid in various organs for short periods of time and these tissues will appear abnormal (see Section 2.3.2.2). Even though these tissues contain large amounts of hydroxyethyl starch, tests of organ function remain normal under these conditions (see Section 3.9).

For obvious ethical reasons, chronic toxicity testing in man has not been undertaken. However, in patients with chronic myelogenous leukaemia undergoing intensive leucapheresis (see Chapter 5 for details) to remove large numbers of granulocytes, a unique situation has been inadvertently created in which to study the effects of the administration of large doses of HES 450/0.70. During each leucapheresis procedure, 30 g HES 450/0.70 were given to each patient to facilitate the separation of granulocytes from the erythrocyte mass. In thirteen

TABLE 3.2 *Chronic toxicity testing in animals treated intravenously with hydroxyethyl starch*

Test animal	Species of hydroxyethyl starch	Dose (g/kg per day)	Frequency	Survival (%)	Reference
Rabbits	HES 81/0.60	0.1	Three months	100	Wiedersheim (1957)
Rabbits	HES 40/0.55	1.8	Four days	100	Irikura *et al.* (1972*q*)
Rats	HES 40/0.55	1.5 or 3.0	Four days	100	Irikura *et al.* (1972*q*)
Rabbits	HES 450/0.55*	0.7	One month	100	Irikura *et al.* (1972*l*)
Rabbits	HES 40/0.55	0.9	Three months	100	Irikura (1972)
Mice	HES 140/0.65	3 or 5	Three weeks†	100	Lindblad (1970)
Dogs	HES 140/0.65	3	Three weeks†	100	Lindblad (1970)
Dogs	HES 450/0.70	7.8–15 (total)	2–5 days	100	Thompson (1963)
Dogs	HES 450/0.70	14.4–27.0 (total)	12–18 days	100	Thompson (1963)
Rabbits	HES 140/0.65	3	One week	95	Lindblad (1970)
Rabbits	HES 40/0.55	3.6	Three months	89	Irikura (1972)
Rabbits	HES 40/0.55	3.6	One month	87	Irikura *et al.* (1972*g*)
Rabbits	HES 450/0.55‡	0.7 or 2.1	One month	83	Irikura *et al.* (1972*l*)
Rabbits	HES 40/0.55	1.8	One month	83	Irikura *et al.* (1972*g*)
Rabbits	HES 40/0.55	1.8	Three months	78	Irikura (1972)
Rabbits	HES 140/0.65	3	Three weeks	50	Lindblad (1970)
Rabbits	HES 450/0.55‡	6.3	One month	17	Irikura *et al.* (1972*l*)
Rabbits	HES 450/0.70	3	Three weeks†	17	Hölscher (1972)
Rabbits	HES 450/0.55*	6.3	One month	0	Irikura *et al.* (1972*l*)

* Suspended in 0.9 per cent sodium chloride
† Animals were injected five times weekly
‡ Suspended in Ringer's lactate

patients treated in this manner, no toxic manifestations were observed after 780-3 750 g HES 450/0.70 had been administered over a period of up to 801 days (see Table 3.6).

3.3. TERATOGENIC EFFECTS

BD-strain rats and Swiss mice administered HES 450/0.70 and tested for teratogenic activity showed no adverse effects (Ivankovic and Bülow 1975). However, high doses (50 g/kg per day) of HES 450/0.70 given intraperitoneally lead to abortion in pregnant rats.

Pregnant mice were administered HES 40/0.55 intravenously and compared with animals given Ringer's lactate (Irikura *et al.* 1972*h*). In pregnant mice given either 1.2 or 3.6 g/kg HES 40/0.55 on days 7-14 of the gestation period, body weight increased in a similar fashion to mice injected with 60 ml/kg Ringer's lactate. In these same pregnant mice, the number of total implants, reabsorbed embryos, and dead and alive foetuses were examined on the 18th day of gestation. In all the categories listed above, there were no significant differences between the three study groups. The body weights of the alive foetuses were also affected in a similar manner in all three groups. Malformations and skeletal deformities of the foetuses in all three groups were similar in number, even though deformities in the 14th rib of foetuses of pregnant mice given 3.6 g/kg HES 40/0.55 were significantly higher compared with pregnant mice administered Ringer's lactate. The number of offspring at birth, the ear-open rate on the 5th and 15th day of parturition, and the rate of weaning were all similar in the three treatment groups. 7-28 days after parturition, the mean body weight of offspring born of pregnant mice given either dose of HES 40/0.55 was significantly higher when compared with that of pregnant mice administered Ringer's lactate. The organ weights of offspring four weeks after parturition of pregnant mice given HES 40/0.55 were usually significantly higher than those of offspring in which the pregnant mice received Ringer's lactate.

Pregnant rabbits were given HES 40/0.55 intravenously and compared with animals given either Ringer's lactate or dextran 70 (Irikura *et al.* 1972*h*). In pregnant rabbits administered either 1.2 or 4.5 g/kg HES 40/0.55 on days 8-18 of the gestation period, the gain in body weight was similar to rabbits given Ringer's lactate (60 ml/kg) but more than rabbits receiving dextran 70 (4.5 g/kg). In these same pregnant rabbits, the number of total implants, and dead and alive foetuses were similar among all four treatment groups on day 28 of gestation. However, the number of retained placentas was higher in rabbits given 4.5 g/kg of either HES 40/0.55 or dextran 70. The number of foetuses born alive, the body weight, structure, and the head and tail length were similar in the pregnant rabbits given either 1.2 or 4.5 g/kg of HES 40/0.55 or Ringer's lactate. Live foetuses born of rabbits given dextran 70 were significantly smaller by the categories listed above. Malformations and skeletal deformities of rabbit foetuses on the 28th day of gestation were similar in all treatment groups, with the exception that imperfect osteogenesis of the 6th rib of the sternum was found in foetuses of pregnant rabbits given 4.5 g/kg of either HES 40/0.55 or dextran 70.

3.4. THE EFFECTS OF ACUTE ADMINISTRATION ON NEUROLOGICAL FUNCTIONS

The intravenous injection (0.6 g/kg) of either HES 40/0.55 or HES 450/0.70 had no effect on the spontaneous electroencephalogram (EEG) (Horii *et al.* 1971; Irikura *et al.* 1972*o*) or monosynaptic reflex of the gastrocnemius nerve of the cat (Irikura *et al.* 1972*o*). Under these same conditions, contraction and the electromyogram (EMG) of the gastrocnemius muscle were not influenced. The intravenous injection of 0.6 g/kg HES 450/0.70 had no effect on the twitch tension of the gastrocnemius muscle in urethane anaesthetised rats (Horii *et al.* 1971).

Spontaneous motor activity (SMA) in mice was not affected after the intravenous injection of 0.6 g/kg HES 450/0.70 (Horii *et al.* 1971). The SMA was depressed, however, in chlorpromazine (0.5 mg/kg s.c.) pretreated mice given 0.6 g/kg HES 450/0.70 (Horii *et al.* 1971) or mice given 6 g/kg HES 40/0.55 (Irikura *et al.* 1972*o*). In the latter case, animals recovered normal SMA after 30 min. Decrease of SMA was observed in mice after the intravenous administration of HES 40/0.55 at dosages approaching lethal values (Irikura *et al.* 1972*n*).

Thiopental sodium sleeping time in mice was unaffected after the injection of either 3 or 6 g/kg HES 40/0.55 (Irikura *et al.* 1972*o*). In the pentetrazole-seizure and strychnine-seizure tests, HES 40/0.55 (3 or 6 g/kg) showed no anticonvulsant effect in mice (Irikura *et al.* 1972*o*). In mice, the intravenous injection of 6 g/kg HES 40/0.55 markedly suppressed the acetic acid writhing, which was considered not as an analgesic effect but due to the dilution of acetic acid with the colloid in the peritoneal cavity (Irikura *et al.* 1972*o*).

Gastrointestinal propulsion of barium sulphate in mice was not affected by the injection of 3 g/kg HES 40/0.55. If the dose was doubled, however, HES 40/0.55 facilitated propulsions (Irikura *et al.* 1972*o*). The oxytocin-induced uterus movement and gastric motility of rabbits *in situ* were not affected by the injection of 1.8 g/kg HES 40/0.55 (Irikura *et al.* 1972*o*).

Spinal discharge in urethane-chloralose anaesthetized cats was not affected by the intravenous injection of 0.6 g/kg HES 450/0.70 (Horii *et al.* 1971). The contractile responses of the nictitating membrane to preganglionic superior cervical nerve stimulation in urethane-chloralose anaesthetized cats were not depressed by the intravenous injection of 0.6 g/kg HES 450/0.70 (Horii *et al.* 1971). The contractile responses of the nictitating membrane to intravenous noradrenaline administration in urethane anaesthetized cats was not suppressed by the intravenous injection of 0.6 g/kg HES 450/0.70 (Horii *et al.* 1971).

Bile secretion in bile-duct cannulated rats or gastric acid or pepsin secretion in pylorus-ligated rats was not affected by the infusion of 1.8 g/kg HES 40/0.55 (Irikura *et al.* 1972*o*).

Histamine, acetylcholine, and 5-HT activities on the guinea-pig small intestine, norepinephrine activity on the rat vas deferens, or acetylcholine and oxytocin activities on the rat uterus were not affected by 10^{-2} g/ml HES 40/0.55 (Irikura *et al.* 1972*o*).

The effect of HES 40/0.55 on the EEG, monosynaptic reflex (MSR), and EMG was further studied in cats immobilized with gallamine triethiodide or anaesthetized with sodium pentobarbitone (Irikura *et al.* 1972*c*). After massive haemorrhage (50-60 ml/kg withdrawal of whole blood), the EEG pattern displayed fast waves of low voltage and spindle bursts; marked flattening of electrical activity was observed in some animals. Decrease in the amplitude of the MSR was also noted. These changes in the EEG, MSR, and EMG were normalised after an equivolume injection of HES 40/0.55.

3.5. HYDROXYETHYL STARCH-INDUCED HYPERAMYLASAEMIA AND MACROAMYLASAEMIA

3.5.1. OBSERVATIONS IN ANIMALS

An incremental rise in the activity of α-amylase in blood occurred in splenectomized dogs after exchange transfusion with HES 450/0.70 to a final haematocrit of 10 per cent (Jesch *et al.* 1975). The highest level of activity occurred three days after transfusion, returning to normal after nine days. In rabbits dosed with 1.8 g HES 40/0.55/kg per day for four consecutive days, the blood activity of α-amylase rose to its highest value 6 hours after the last injection, falling to basal levels after 24 hours (Irikura *et al.* 1972*p*).

3.5.2. OBSERVATIONS IN MAN

The activity of α-amylase in blood begins to rise above basal levels 2-6 hours after a single infusion of 30 g of either HES 40/0.55 (Adachi *et al.* 1972; Gofferje and Hosslick 1977), HES 264/0.43 (Mishler and Dürr 1979), HES 350/0.60 (Mishler and Dürr 1980), or HES 450/0.70 (Boon *et al.* 1976; Dürr *et al.* 1978; Gofferje and Hosslick 1977; Köhler *et al.* 1976, 1977*a, b, e*; Mishler *et al.* 1977*a*). Higher than normal activity is not observed before this period, even when up to 45 g are given (Janes *et al.* 1977). Maximum activity in blood is reached 12-24 hours after injection of any species of hydroxyethyl starch. The overall persistence of this induced hyperamylasaemia is approximately directly related to the intravascular half-life of the injected species (Table 3.3).

In patients dosed twice (2 × 30 g) within 24 hours with either HES 40/0.55 or HES 450/0.70, the highest activity of α-amylase in blood was reached 48 hours after the last injection, returning to normal limits in 4-7 days (Gofferje and Hosslick 1977). In normal subjects given 30 g HES 450/0.70 on three consecutive days (total 90 g), the highest activity of α-amylase occurred 12 hours after the third injection and returned to basal levels after 312 hours (Mishler *et al.* 1977*a*).

TABLE 3.3 *Duration of hyperamylasaemia as related approximately to the intravascular half-life of various species of hydroxyethyl starch*

Species of hydroxyethyl starch	Intravascular half-life (hours)	Appearance of increased α-amylase activities after injection (hours): Highest level	Return to basal level	Reference
HES 40/0.55	2–4	24	96	Gofferje and Hosslick (1977)
HES 264/0.43	3–5	12	120	Mishler and Dürr (1979)
HES 350/0.60	11–13	24	192	Mishler and Dürr (1980)
HES 450/0.70	24–36	24	168	Boon *et al.* (1976) Gofferje and Hosslick (1977) Köhler *et al.* (1977*a,b,e*)

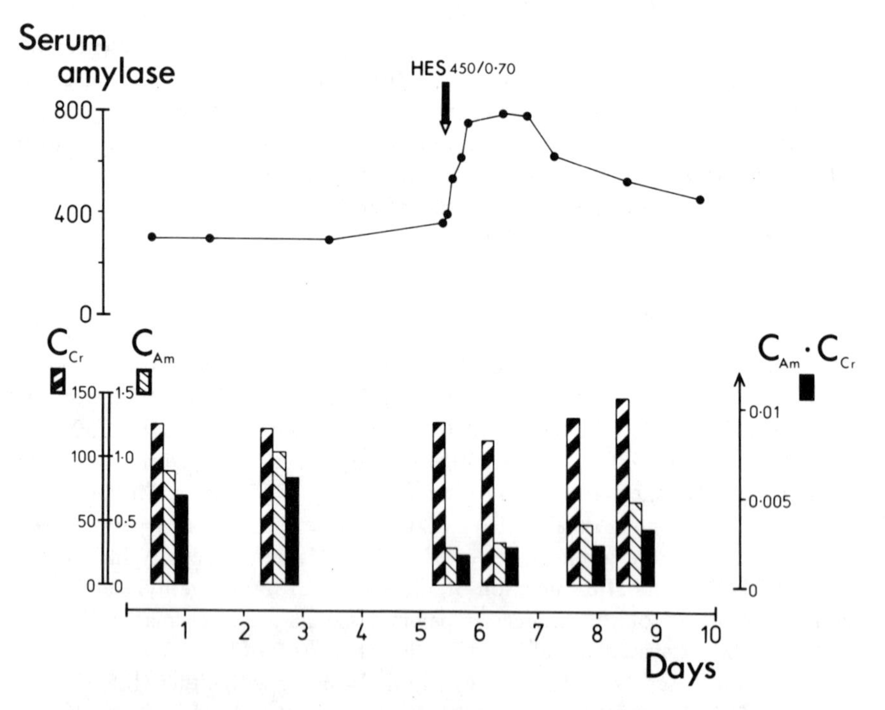

Fig. 3.1. (Upper) Changes in the concentration (U/l) of α-amylase in the blood of one male test subject after he was given 30 g HES 450/0.70. (Lower) C_{cr} and C_{am} (ml/min), and the ratio C_{am}/C_{cr}, were determined in the same subject, before and after the injection. As the concentration of α-amylase in blood rose markedly, the C_{am} diminished even though C_{cr} remained at approximately preinfusion levels. (From Mishler and Dürr 1980.)

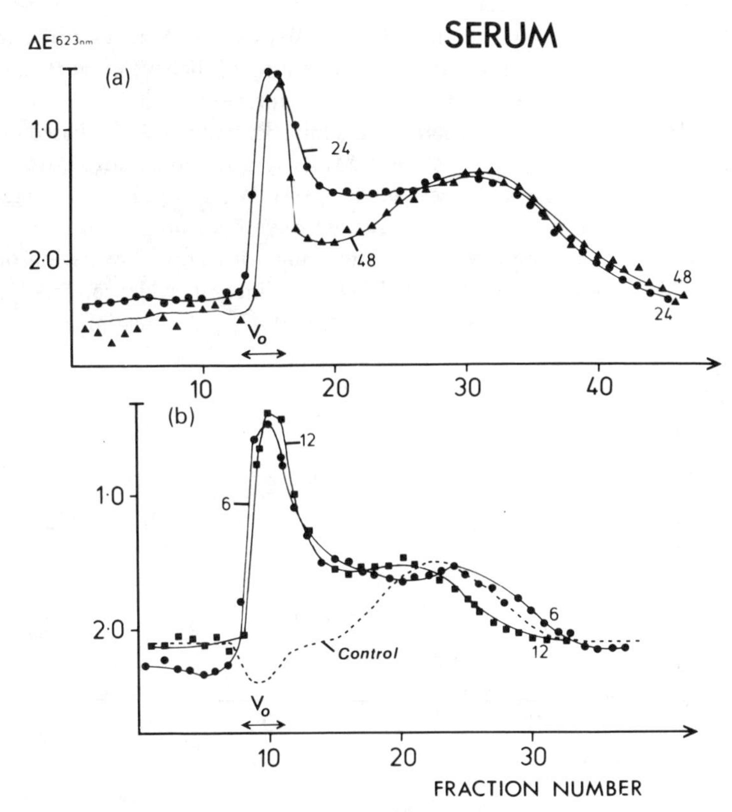

Fig. 3.2. Gel filtration elution patterns of α-amylase activity in aliquots of serum recovered from one male subject for up to 48 hours after he was given HES 350/0.60. The void volume (V_0) as determined by blue dextran was eluted together with fractions 8–11 in those chromatograms displayed in A (control and 6 and 12 hours after injection) and (because of altered volume of drops) together with fractions 13–16 in those chromatograms displayed in B (24 and 48 hours after injection). Absorption was low in those fractions not having enzyme activity, since the intensity of the blue starch–iodine colour is diminished by starch digestion. The bulk of serum proteins were eluted together with the V_0 and were responsible for the slightly higher absorption in fractions 8–11 of the control chromatogram (protein interferes significantly with the starch–iodine reaction). By 48 hours after infusion, the elution pattern was returning to normal; there was evidence of a progressive disassociation of the hydroxyethyl starch–amylase complex, and this was paralleled by the fall in the concentration of α-amylase in serum (refer to Fig. 3.3). (From Mishler and Dürr 1980.)

3.5.3. MECHANISM AND MACROAMYLASE FORMATION

The hyperamylasaemia observed after the infusion of various molecular weight species of hydroxyethyl starch could be caused by any of the following: 1) increased release of enzyme from damaged sites of synthesis or storage; 2) decreased excretion or catabolism; 3) formation of macroamylase complexes or

4) a combination of several of these factors. With the techniques available for monitoring renal, hepatic, and pancreatic function, it is believed that temporary damage to sites of synthesis or storage of α-amylase would not explain the hydroxyethyl starch-induced hyperamylasaemia. This conclusion is based on the fact that tests of organ function are not adversely affected by large intravenous doses of hydroxyethyl starch (see Section 3.9). What appears more likely is that the large glycogen-like structure of hydroxyethyl starch forms a complex with α-amylase (macroamylase), thus interfering with normal excretion of the enzyme. This hypothesis is supported by two lines of evidence. Firstly, the clearance of α-amylase (C_{am}) and the ratio of C_{am} to creatinine clearance

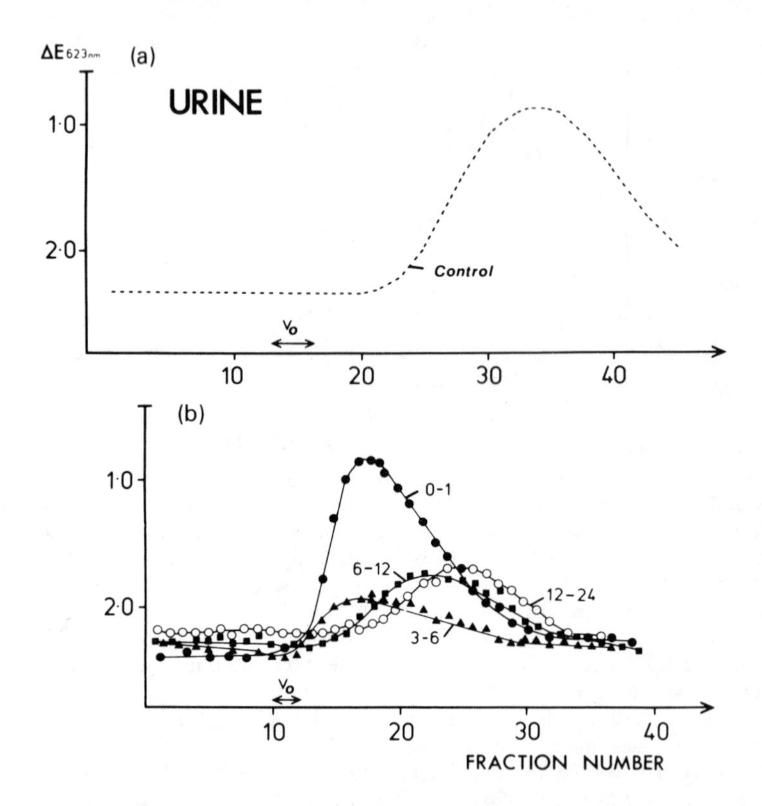

Fig. 3.3. Gel filtration elution profiles of α-amylase activity in aliquots of urine collected from a male subject (see Fig. 3.2) up to 24 hours after he was given HES 350/0.60. The V_0 as determined by blue dextran was eluted with fractions 10–12 in those chromatograms displayed in B (recovered aliquots of urine taken 0–24 hours after dosing), and (because of altered volume of drops) together with fractions 13–16 in the control chromatogram shown in A. By 24 hours after injection, the elution profile was returning to normal, and this was seen concomitantly with an increase in the rate of excretion of α-amylase. (From Mishler and Dürr 1980.)

(C_{cr}), as well as the excretion of α-amylase, are all markedly reduced after the injection of any species of hydroxyethyl starch (Dürr *et al.* 1978; Gofferje and Hosslick 1977; Köhler *et al.* 1977*a*; Mishler and Dürr 1979, 1980; Watzek *et al.* 1978) (Fig. 3.1), even although α-amylase activity in duodenal secretion is not affected (Köhler *et al.* 1977*e*).

Secondly, macroamylasaemia is observed after the infusion of hydroxyethyl starch (Dürr *et al.* 1978; Kirch *et al.* 1978; Köhler *et al.* 1977*a, b, e*; Mishler and Dürr 1979, 1980). This induced macroamylase is apparently of a molecular weight or molecular radius that does not allow excretion through the glomerulus, thus leading to hyperamylasaemia. The formation of this macroamylase complex appears to be a unique property of hydroxyethyl starch, since dextran 40 or gelatin are unable to induce such colloid–enzyme aggregates (Köhler *et al.* 1977*b*). The elution profile of α-amylase activity in blood has shown that the induced hydroxyethyl starch–enzyme complex is significantly altered with time (Fig. 3.2). These observed changes can be explained by the enzymatic hydrolysis of hydroxyethyl starch, the splitting of macroamylase complexes, and the increase with time of the number of low molecular weight complexes. A portion of these low molecular weight complexes appear to be excreted in urine (Fig. 3.3).

There is probably a slow progressive disassociation of these complexes in blood, accompanied by release of hydroxyethyl starch and α-amylase. This would explain the continual movement of the macroamylase elution profile towards normal patterns in aliquots of blood and urine recovered after dosing (Figs. 3.2 and 3.3). Also, although this has not been substantiated, the site(s) of binding of α-amylase to hydroxyethyl starch used in the macroamylase complex is probably distinctly different from the active enzymatic site.

The molecular weight distribution of naturally occurring macroamylase is distinctly different from, and possibly more homogeneous than, that of the hydroxyethyl starch-induced macroamylase complexes (Bode *et al.* 1977; Dürr *et al.* 1978). Naturally-occurring macroamylase may result from complex-formation between α-amylase and various macromolecular components, such as immunoglobulins, polysaccharides, and glycoproteins. In conclusion, the macro-amylasaemia induced by hydroxyethyl starch represents the non-pathological formation of large colloid–enzyme complexes not easily filtered at the renal glomerulus, and should be differentiated clinically from disease-induced or drug-induced hyperamylasaemia.

3.6. HISTAMINE RELEASE, IMMUNOGENICITY, AND THE INCIDENCE OF 'ANAPHYLACTOID-TYPE' REACTIONS

The anaphylactoid reaction of the rat to dextran is highly characteristic and has been described as consisting of oedema and erythaema of paws, snout, and genitalia (Walton *et al.* 1966). It is also generally recognized that the dog and other members of the canine family exhibit distinct hypotensive reactions after intravenous injection of certain macromolecular polymers and that these reactions have the characteristics of a histamine-release response (Walton *et al.* 1959). The incidence and severity of 'anaphylactoid-type' reactions in animals and in man, the release of histamine induced by the infusion of hydroxyethyl starch, and the immunogenicity of this molecule is discussed below.

3.6.1. STUDIES IN ANIMALS

3.6.1.1. RELEASE OF HISTAMINE

When HES 40/0.55, dextran 70, or Ringer's lactate are incubated *in vitro* with an equal volume of heparinized rat or rabbit blood for one hour at 37 °C, no significant increases of the blood histamine concentration were observed for either HES 40/0.55 or Ringer's lactate (Irikura *et al.* 1972*i*). However, rabbit blood incubated with dextran 70 had significantly high concentrations of histamine. Incubation of dog blood with dextran 70 caused elaboration of a vasodilating principle inhibited by antihistamines, but this was not formed by HES 450/0.70 (Silk 1966).

Administration of either 1.2 or 6 g/kg HES 450/0.70 to dogs evoked no hypotensive reactions and produced no increase in arterial plasma histamine concentrations (Murphy 1965*a*, *b*; Murphy *et al.* 1965; Thompson and Walton 1964*a*). With more sensitive methods for determining histamine concentrations in blood, it was shown that HES 450/0.70 increased blood histamine concentrations in dogs, but only with a low incidence and only up to concentrations which did not influence the blood pressure (Lorenz and Doenicke 1978). In dogs given starch (2.1 g/kg) that had not been hydroxyethylated, no hypotensive reactions were observed, even although slight amounts of histamine were released after intravenous injection (Walton *et al.* 1959).

3.6.1.2. INCIDENCE OF 'ANAPHYLACTOID-TYPE' REACTIONS

No anaphylactoid reactions were observed in rats injected intraperitoneally with 0.1–0.9 g/kg HES 450/0.70 or 0.1–0.3 g/kg HES 150/0.70 (Walton *et al.* 1966). White mice injected intraperitoneally with HES 450/0.70 in doses of 0.3, 1.5, or 3 g/kg were without evidence of anaphylactoid reactions. Paw thickness after injection was not significantly altered in rats given either 0.1 or 0.9 g/kg HES 450/0.70 or 0.1 or 0.3 g/kg HES 150/0.70 intraperitoneally with an agar plus hydroxyethyl starch admixture. After the intraperitoneal injection of 0.3 g/kg dextran 75 into rats, these animals developed erythema and oedema of the paws, nose, ears, eyelids, and tongue accompanied by generalized pruritis (Silk 1966). No such effects were noted after a similar injection of HES 450/0.70.

Vascular permeability and oedema of the paw in rats and mice were unaffected 2 hours after the intravenous injection of 0.3 g/kg HES 40/0.55 (Irikura *et al.* 1972*i*). Increased hind-paw oedema was observed temporarily in rats treated with 18 mg/ml HES 40/0.55, but dextran 70 caused pronounced oedema at a dose as low as 1 mg/ml (Irikura *et al.* 1972*m*). No effect on vascular permeability was seen in rats given HES 40/0.55. Anaphylactoid purpura reactions were not observed in rats or mice injected intravenously with agar and HES 40/0.55 (Irikura *et al.* 1972*i*).

In dogs injected intravenously with HES 450/0.70, anaphylactoid reactions occurred but to a lesser extent than with either dextran or gelatin (Lorenz and Doenicke 1978). Passive cutaneous anaphylaxis (PCA) or Arthus reactions did not occur in guinea-pigs given either HES 40/0.55 (Irikura *et al.* 1972*j*) or HES 450/0.70 (Maurer 1965; Maurer and Berardinelli 1968). No PCA or anaphylactoid reactions were observed in rabbits given HES 40/0.55 (Irikura *et al.* 1972*j*). After the subcutaneous injection of HES 40/0.55 with Freund's complete adjuvant to mice, 10–11 per cent of mice displayed positive PCA reactions (Irikura *et al.* 1972*j*). With Freund's incomplete adjuvant, HES 40/0.55 did not induce PCA in mice. Without adjuvant, however, HES 40/0.55 induced one out of twenty and dextran 190 induced two out of twenty cases of PCA. The antiserum to dextran 190 displayed cross-reactivity to HES 40/0.55 in a guinea-pig PCA test.

3.6.1.3. SCHWARTZMAN-TYPE TISSUE REACTIONS

Severe bleeding or tissue necrosis will occur in rabbits injected subcutaneously with a filtrate of *B. typhosus* (preparative injection) followed 24 hours later with an intravenous injection (provocative injection) of the same filtrate. This phenomenon has been termed the Schwartzman reaction. With this scheme, local macroscopic reactions have occurred in rabbits when antibiotics are used as the preparative injection and hydroxyethyl starch as the provocative substance (Goto and Kimura 1976; Goto *et al.* 1972). Hyperaemic changes occur with penicillin, chloramphenicol, and polymixin-B and HES 450/0.70; kanamycin, erythromycin, tetracycline, chloramphenicol, and polymixin-B and HES 150/0.70; and lincomycin, cephalosporin, erythromycin, tetracycline, chloramphenicol, and polymixin-B and HES 40/0.55 (Goto and Kimura 1976). In the instance where either HES 450/0.70 or HES 150/0.70 is used as the preparative injection, there are no reactions when numerous antibiotics are used as the provocative substance (Goto *et al.* 1972).

3.6.1.4. IMMUNOGENICITY

Immune responses against HES 450/0.70 or hydroxyethyl starch preparations possessing a M_W of 2 500 000 did not occur in either rabbits or guinea-pigs as tested by the precipitin reaction or agar diffusion (Maurer and Berardinelli 1968). Additional studies in rabbits and mice have shown HES 40/0.55 and HES 450/0.70 not to be immunogenic (Kimura *et al.* 1971; Matsumoto *et al.* 1971; Roberts 1965; Roberts and Pagones 1965). Hydroxyethyl starch has been shown not to be immunogenic in dogs or cats (Wiedersheim 1957). In rabbits either injected subcutaneously with HES 40/0.55 and Freund's complete adjuvant or injected intravenously with HES 40/0.55, antibody was not detected by the precipitin reaction, gel diffusion or sensitized red-cell haemagglutination (Irikura *et al.* 1972*j*; Kanehako 1972).

Rabbits immunized with a hydroxyethyl starch-bovine albumin conjugate have, however, produced antibodies against this substance as demonstrated by positive gel diffusion and passive haemagglutination tests (Richter and de Belder 1976). Even though hydroxyethyl starch is not conjugated to albumin in nature, the significance of these findings is not fully understood.

3.6.2. STUDIES IN HUMANS

3.6.2.1. RELEASE OF HISTAMINE

The concentration of histamine in heparinized human blood, mixed *in vitro* with an equal volume of HES 40/0.55 did not increase significantly after incubation at 37 °C for one hour (Irikura *et al.* 1972*i*).

In male volunteers, intravenously infused weekly for five weeks with HES 450/0.70 sensitivity to HES 450/0.70 was assessed by the release of histamine from their leucocytes up to three months after the final infusion (Brickman *et al.* 1966*a*,*b*). The percentage of histamine released from leucocytes challenged with HES 450/0.70 was less than 5 per cent except at one month where 3/7 values were greater than 5 per cent. The maximum of histamine released from leucocytes challenged with HES 450/0.70 was under 10 per cent, and most values were under 5 per cent. Reactions producing values of histamine release under 10 per cent are probably non-specific and those giving values under 5 per cent are either non-specific or within experimental error.

No significant release of histamine was observed in ten subjects after the rapid injection of HES 450/0.70 (Lorenz *et al.* 1975; Lorenz 1975). The actual increase in plasma histamine concentrations in two out of ten subjects was less than 3.0 ng/ml (Lorenz and Doenicke 1978; Lorenz *et al.* 1978). The increase in plasma histamine concentrations in these two subjects occurred surprisingly late after infusion (between 20 minutes and one hour).

3.6.2.2. INCIDENCE OF 'ANAPHYLACTOID-TYPE' REACTIONS

In ten normal volunteers given a rapid infusion of HES 450/0.70, no allergoid or anaphylactoid reactions were observed (Lorenz and Doenicke 1978; Lorenz *et al.* 1978). One case of an 'anaphylactoid-type' reaction after the infusion of a small quantity of HES 450/0.70 has been reported (von Matthiesen *et al.* 1977/78). Three pyrogenic reactions and two mild urticarial reactions were observed in 36 patients given a single 60 g injection of HES 450/0.70 in early clinical trails (Metcalf *et al.* 1970). Of thirty-one patients given HES 450/0.70, one patient developed urticaria (Amakata *et al.* 1972*a*). Out of a total of 779 patients receiving either HES 450/0.55 alone or HES 40/0.55 and electrolyte solutions, one patient was reported to have flushing accompanied with a slight chill (Kori-Lindner and Hubert 1978) (Table 3.4). In a total of 10 273 patients receiving HES 450/0.70, eight patients experienced an 'anaphylactoid-type' reaction (Ring *et al.* 1976). In three of these eight patients, the untoward effect was restricted to the site of infusion or skin reactions. Five patients showed marked

TABLE 3.4 *The incidence of anaphylactoid-type reactions in patients receiving hydroxyethyl starch*

Species of hydroxyethyl starch	Total number of patients	Total number of all reactions	Frequency (%)	Reference
HES 450/0.70	150	4	2.7	Schöning and Koch (1975)
HES 40/0.55	779	1	0.13	Kori-Lindner and Hubert (1978)
HES 450/0.70	5 780	0	0.00	Messmer *et al.* (1978)
HES 450/0.70	10 273	8	0.08	Ring *et al.* (1976)
HES 450/0.70	16 405	14	0.09	Ring and Messmer (1977*a,b*); Messmer *et al.* (1978)
HES 450/0.70	550 350	26	0.005	Beez and Dieti (1979)
Total (HES 450/0.70 only)	582 958	52	0.009	

haemodynamic changes, tachycardia was observed in six patients, and hypotension was noted in four patients. One patient developed anaphylactic shock after infusion of 10 ml (600 mg) HES 450/0.70 but recovered after treatment with 500 mg prednisolone.

The incidence of 'histamine-like' reactions was evaluated in 750 randomized patients given gelatin, dextran, or HES 450/0.70 (Schöning and Koch 1975). Each patient was infused with 500 ml of the plasma expander at a rate of 25-30 ml/min. There was a positive correlation between the frequency of untoward reactions and the substance used; none, however, was related to age or general risk of surgery. Untoward reactions were observed in 21.3 per cent with derivatives of gelatin as compared with 4.7 per cent with dextran and 2.7 per cent with HES 450/0.70. Untoward effects occurring in the group of patients receiving HES 450/0.70 consisted only of local reactions.

In a large multicentre prospective trial of the incidence of reactions after the infusion of commonly used colloid volume substitutes, it was reported that the incidence of *severe* reactions (shock, cardiac and/or respiratory arrest) was 0.002 per cent for dextran 40, 0.003 per cent for plasma-protein solutions, 0.006 per cent for HES 450/0.70, 0.02 per cent for dextran 60/75, and 0.04 per cent for gelatin solutions (Ring and Messmer 1977*a*, *b*; Messmer *et al.* 1978). In fourteen out of 16 405 patients (incidence 0.09 per cent) receiving HES 450/0.70 and experiencing an 'anaphylactoid-type' reaction, five patients had local skin reactions and/or mild fever. Eight patients had a measurable but not life-threatening cardiovascular reaction (e.g. tachycardia or hyptension) or a gastrointestinal (nausea) or respiratory disturbance. One patient experienced shock.

In a large retrospective analysis of the incidence of 'anaphylactoid-type' reactions in patients given either HES 450/0.70 or dextran 75, the incidence of severe untoward effects (shock, cardiac and/or respiratory arrest) was 0.0041 per cent for dextran 75 and 0.0004 per cent for HES 450/0.70 (Beez and Dietl 1979). In twenty-six out of 550 350 patients (frequency 0.005 per cent) receiving HES 450/0.70 and experiencing an untoward effect, six patients had a local skin reaction and/or mild fever. Eighteen patients had a measurable but not life-threatening cardiovascular reaction (e.g. tachycardia or hypotension) or gastrointestinal (nausea) or respiratory disturbance. One patient experienced shock and one patient had cardiac and/or respiratory arrest.

In a large multicentre retrospective study of normal volunteers undergoing centrifugal leucapheresis (see Chapter 5 for details of the procedure) in which HES 450/0.70 was added to the input-line of the continuous- or intermittent-flow centrifuge, no 'anaphylactoid-type' reactions were observed, even in some volunteers given hydroxyethyl starch on multiple occasions (Table 3.5. Mishler 1975). In a group of fifty-five patients diagnosed as having chronic myelogenous leukaemia and undergoing treatment with the continuous-flow centrifuge to which HES 450/0.70 was added to the input-line, no 'anaphylactoid-type' reactions were observed, even in patients receiving more than 3000 g hydroxyethyl starch over 1.3-2.2 years (Table 3.6; Mishler 1975).

TABLE 3.5 *Numbers of normal donors undergoing single and multiple donations and additional agents**

Treatment	Procedures experienced										
	1	2	3	4	5	6	7	8	9	10	11
Hydroxyethyl starch†	149	23	14	5	1	2	0	0	1	0	0
Hydroxyethyl starch† and dexamethasone	32	6	3	2	0	1	1	0	0	0	0
Hydroxyethyl starch† and etiocholanolone	61	25	84	15	14	2	3	1	1	1	2
Total donors	242	54	101	22	15	5	4	1	2	1	2

* From Mishler (1975)
† HES 450/0.70
‡ A few donors received both treatment regimens

TABLE 3.6 *Numbers of patients with chronic myelogenous leukaemia undergoing leucapheresis, according to numbers of procedures experienced**

Treatment	Procedures experienced 2–10	11–25	26–50	51–75	76–100	101–125
Hydroxyethyl starch†	28	14	5	3	2	3‡

* From Mishler (1975)
† HES 450/0.70
‡ Days of study ranged from 472 to 801

3.6.2.3. IMMUNOGENICITY

No precipitating antibodies could be detected in normal volunteers up to three months after receiving large intravenous injections of HES 450/0.70 (Brickman *et al.* 1966*a*,*b*) or after immunization with milligram doses of this same species of hydroxyethyl starch (Maurer and Berardinelli 1968).

No precipitating anti-hydroxyethyl starch antibodies were detectable in the sera of eight patients experiencing 'anaphylactoid-type' reactions after the infusion of HES 450/0.70 (Ring *et al.* 1976). In five out of these eight patients, skin tests with HES 450/0.70 in different dilutions were positive (immediate-type reactions) ten days to two months after the untoward reaction. In seven out of these eight patients, the concentrations of serum immunoglobulins were within normal limits after the infusion. The concentrations of complement factors decreased during the infusion of HES 450/0.70 in all eight patients: most pronounced were C3 in two out of eight patients, C4 in one out of eight, and C3-activator in two out of eight. Lymphocyte transformation tests were conducted in five out of these eight patients. Lymphocyte (T- and B-cell) function was normal, as measured by phytohaemagglutinin and pokeweed mitogen stimulation.

With the more sensitive technique of passive haemagglutination, preformed hydroxyethyl starch-reactive antibodies of low titre were detected in 26 per cent of sera of 268 normal volunteers (Richter *et al.* 1977). The highest titres, comprising only 1 per cent of the sera tested, ranged from 1:4 to 1:8. Comparing the occurrence of preformed antibodies in man reactive with plasma substitutes, there is evidence that both titres and frequency are highest for derivatives of gelatin, intermediate for dextran, and lowest for HES 450/0.70. The presence of preformed antibodies to plasma colloid substitutes has also been examined in an additional study (Irikura *et al.* 1972*j*). With the PCA test in guinea-pigs, antibodies to dextran 70 were detected in three out of 106 normal volunteers tested (frequency 2.8 per cent), while one volunteer in this same group had a weak antibody to HES 40/0.55 (frequency 0.9 per cent). Recently, passive haemagglutination tests have been performed on the sera of 75 random normal donors (Lalezari, unpublished observations). Fifty-five of these sera were tested against hydroxyethyl starch-coated cells alone, and 20 were tested against both hydroxyethyl starch and dextran-coated cells. Eleven of these donors showed a 1+ to 2+ positive reaction against dextran-coated cells, but none had activity against hydroxyethyl starch. The passive haemagglutination test used primarily detects 19S antibodies.

3.6.3. CONCLUSIONS

The results of several large multicentre studies indicate that patients receiving synthetic volaemic colloids may show anaphylactoid reactions (Beez and Dietl 1979; Ring and Messmer 1977*a*; Rittmeyer 1976; Seidemann 1979). However, in terms of the frequency and severity of these reactions, hydroxyethyl starch should be considered 'safer' than most other synthetic colloids used for augmentation of the plasma volume (Rittmeyer 1976). Little is known about the mechanism underlying these 'anaphylactoid-type' or 'histamine-like' reactions to hydroxyethyl starch. It has been postulated that the mechanism may include complement activation (Doenicke *et al.* 1977; Richter *et al.* 1978; Ring *et al.* 1976; Ring and Messmer 1977*a*). Studies conducted in animals and man show that hydroxyethyl starch does not elicit antibody formation or release clinically significant quantities of histamine. However, some individuals in the general population may possess preformed antibodies against substances in plasma that cross-react with hydroxyethyl starch.

3.7. RHEOLOGICAL EFFECTS

In this section I will focus on the interaction of hydroxyethyl starch and the formed elements of blood, especially the erythrocyte. The main emphasis will be directed towards the 'membrane coating' induced by this colloid and the subsequent effect on rouleaux formation, migration in an electrical field, and – more importantly for the clinician – the effect on blood typing and cross-matching.

3.7.1. ROULEAUX FORMATION AND SUSPENSION STABILITY

The plasma protein coat of erythrocyles, known to consist mainly of fibrinogen and gamma globulin, is only slightly increased by the addition of hydroxyethyl starch (Thompson 1965, 1966; Goto and Aochi 1973, 1974; Goto *et al.* 1972*a*). Electrophoretic analyses suggest that hydroxyethyl starch does not interact with plasma proteins, and that changes in red-cell charge and suspension stability may be due to alterations in the structure of the erythrocyte envelope, because changes in protein content alone are not sufficient to account for alterations in suspension stability (Thompson 1966).

What appears to be true, however, is that the influence of hydroxyethyl starch on red-cell charge and suspension stability is a function of molecular weight (Goto *et al.* 1972*a*; Goto and Aochi 1973, 1974; Matsumoto 1974). For example, the electrophoretic velocity of erythrocytes has been shown to be inversely related to the molecular weight of hydroxyethyl starch (Goto and Aochi 1973, 1974). Red cells suspended (30 per cent haematocrit) in HES 40/0.55 achieve a final velocity of 0.79 mm/min compared with 0.41 and 0.39 mm/min when cells are suspended in HES 150/0.70 and HES 450/0.70, respectively. Enhanced electrophoretic mobility, under these conditions, is correlated with an increase in electronegativity concomitant with a reduction in the overall charge of the red cell.

The mechanism by which high-molecular-weight colloids increase red-cell aggregation and sedimentation is somewhat obscure. The classical experiments performed by Thorsen and Hint (1954) have, however, shown that under normal test conditions the sedimentation of erythrocytes increases exponentially as the concentration of various colloids in the blood is raised. In this regard, two properties of colloids influence the erythrocyte sedimentation rate (ESR) independently. There appears to be a critical molecular weight (C_{mw}) of a given colloid below which sedimentation is not increased. For example, the C_{mw} for dextran is 59 000 and that of gelatin is 18 000 (Thorsen and Hint 1954). In addition to C_{mw}, colloids have a critical concentration (C_c) in the suspending fluid below which the suspension stability is not decreased. The C_c of dextrans with a $M_{\bar{w}}$ of 62 000 and 82 000 are 1.80 and 0.98 g/dl, respectively. Studies *in vitro* and *in vivo* conducted on animal (Hazi *et al.* 1972; Thompson and Walton 1964*b*) and human whole blood (Goto and Aochi 1973; Kitmura 1972; Mishler *et al.* 1977*a*, 1978*a*, 1979*b*; Mishler 1980*b*; Thompson 1966) have documented that the C_{mw} and C_c for hydroxyethyl starch is less than 300 000 and 0.25–0.40 g/dl, respectively. In addition to C_{mw} and C_c, the number of hydroxyethyl groups attached to the starch molecule appears to influence suspension stability. For example, given two species of hydroxyethyl starch of approximately equal molecular weight at a final concentration of 1 per cent in whole blood, red-cell sedimentation is increased more by the species having a higher degree of hydroxyethyl group substitution (Woods *et al.* 1980). At higher concentrations of hydroxyethyl starch in whole blood ($\geqslant$ 1.5 per cent), this effect is less pronounced.

Microscopically, red cells are spread about freely in random positions when suspended in whole blood containing hydroxyethyl starch in which the C_c or C_{mw} has not been surpassed (Goto and Aochi 1973; Janes *et al.* 1977). These red cells may touch each other, but they never stick together or move towards each other. If the concentration and/or molecular weight of hydroxyethyl starch approaches the critical level, red cells coming into contact with one another move into short, loosely-constructed rouleaux formations. In these circumstances, red cells do not adhere to each other but are lightly juxtaposed and can be easily disrupted by the slightest motion. When either the C_c or C_{mw} of hydroxyethyl starch in solution is surpassed, the intervening distances between individual red cell surfaces diminishes. The cells are flattened out, their thickness diminishes, and their diameter increases. With still further increases in the concentration and/or molecular weight of hydroxyethyl starch, the red cells coming into contact with each other no longer shift about so that their flat surfaces come to face each other, but remain joined to each other at whatever place they touch. The masses so formed are quite irregular and 'bricks' of rouleaux are absent.

The formation of a film in colloid solutions, at phase boundaries comparable with the borderline between blood cell surface and plasma, is a well-known phenomenon. The cause of this formation is not yet fully known. The formation of a film appears to be connected with the increased concentration of colloids at the phase boundaries: the tendency to film formation in colloid solutions increases with the concentration of colloid in the solution and the molecular weight of the colloids (i.e. with the same factors found to increase the sedimentation rate). Accordingly, there is a relation between the tendency of colloids to form films at phase boundaries and their ability to provoke aggregation of blood cells. Hence the surface film must be formed from the suspension fluid. It does not consist of any special substance from the blood cell but is formed by the concentration of the colloid on the surface of the blood cell by adsorption. Thus there develops a colloid-rich, gel-like liquid phase in equilibrium with the colloids in the suspension fluid.

This is in entire agreement with the observation that film is formed when the suspension fluid contains a sufficient concentration of hydrophilic colloid with highly asymmetrical, chain-like molecules with a weight above the critical limit. The addition of low-molecular-weight colloid, which reduces the average molecular weight to below the critical level, causes the dissolution of the surface film and the disappearance of the aggregation. The same effect is caused by a decrease in the colloid concentration below the critical point by dilution of the suspension fluid.

The lowest content of colloid in the suspension fluid which causes the formation of a surface film is found to correspond to the critical concentration. The formation is also greatly dependent upon the degree of asymmetry of the colloid molecules and likewise upon their molecular weight distribution.

3.7.2. BLOOD TYPING AND CROSS-MATCHING IN THE PRESENCE OF HYDROXYETHYL STARCH

On occasions where samples of blood are taken for typing and cross-matching determinations *after* infusion of hydroxyethyl starch, one should be aware that because these materials affect rouleaux formation and red-cell surface charge characteristics (*vide supra*), *false-positive* (pseudo-agglutination) reactions may be encountered.

3.7.2.1. *IN VITRO*

Whole-blood concentrations of 20–30 mg/ml HES 40/0.55 generally produce *false-positive* (pseudo-agglutination) results in approximately 20 per cent of tests conducted. A concentration of 7–12 mg/ml in whole blood results in less than 5 per cent macroscopic agglutination (Kox *et al.* 1978). Whole blood diluted 11–50 per cent with HES 40/0.55 did not yield false-positives in the ABO and Rh systems; in addition, the albumin cross-match test was not interfered with, nor was the Bromelin-enzyme test (Kox and Howekamp 1979).

Artificially prepared whole-blood mixtures containing up to 12 mg/ml HES 450/0.70 could be typed and cross-matched normally (Kleine 1975; Kox *et al.* 1978; Kox and Howekamp 1979), even although in approximately 60 per cent of the tests, pseudo-agglutination was observed. Greater than 80 per cent of blood tested under these same conditions gave *false-positive* results when the concentration of HES 450/0.70 was raised to 30 mg/ml. In whole blood dilutions ranging between 11 and 50 per cent, false-positive reactions were observed in the Bromelin-enzyme test (Kox and Howekamp 1979).

3.7.2.2. *IN VIVO*

In patients administered 30 g HES 40/0.55 after the infusion of 1000–2000 ml Ringer's lactate solution, and then typed, no signs of pseudo-agglutination were observed (Goto *et al.* 1972*b*; Goto and Matsumoto 1973; Goto 1974). The typing of blood was not interfered with in patients receiving 44–54 g HES 40/0.55 (Adachi *et al.* 1972).

Determination of blood-group types was not affected in patients given 30 g HES 150/0.70 after the infusion of 1000 ml Ringer's lactate solution (Goto and Matsumoto 1973; Goto 1974). However, in three out of twelve cases, a *false-positive* serological reaction for syphilis was noted.

In eight normal subjects infused on three separate occasions with 15, 30, or 45 g HES 450/0.70, rouleaux formation was observed to be dose-related (Table 3.7), but blood typing and cross-matching studies were normal in all cases (Janes *et al.* 1977). The estimated concentration of HES 450/0.70 in blood ranged between 2.8 and 12 mg/ml. Rouleaux formation was observed in volunteers or patients dosed either with 60 g HES 450/0.70 or 30 g HES 450/0.70 (after the infusion of 1000 ml Ringer's lactate solution), but in all cases the determination of the blood group type was not affected (Ballinger *et al.* 1966*a*, *b*;

Goto and Matsumoto 1973; Goto 1974). A concentration of 12–14 mg/ml HES 450/0.70 in blood is usually achieved by infusion of 60 g HES 450/0.70 (Metcalf *et al.* 1970; Mishler *et al.* 1977*a*).

TABLE 3.7 *The effects in vivo of varying blood concentrations of HES 450/0.70 on rouleaux formation**

Quantity infused (ml)	Calculated concentration in blood (mg/ml)	Rouleaux formation
250	2.8–4.0	None
500	5.8–8.0	Microscopic
750	8.8–12.0	Macroscopic

* From Janes *et al.* 1977

Note

Induced rouleaux formation (pseudo-agglutination) can be easily distinguished from true agglutination by the addition of 0.9 per cent isotonic saline, which readily disperses the former. A 1 per cent solution of either glycine (Koop and Bullit 1945) or sodium salicylate (Inokuchi 1950) abolishes the pseudo-agglutination induced by gelatin and sodium alginate, respectively. Whether either glycine or sodium salicylate would be appropriate for dispersion of the agglutination induced by hydroxyethyl starch remains to be answered.

Caution

During cross-matching studies after the infusion of large volumes of hydroxyethyl starch, weak antibodies may be missed because of a high degree of dilution of the patient's serum. Increasing the sensitivity by enzyme techniques (papain) is not recommended, as irreversible non-specific reactions can be detected at a low degree of volume replacement (Kleine 1978).

3.8. EFFECTS OF COAGULATION, FIBRINOLYSIS, AND HAEMOSTASIS

Despite more than two decades of research, the explanation of the long-known haemostatic failure following the use of some natural and synthetic volaemic colloids remains obscure. In reviewing the literature on this subject, it is necessary to differentiate between changes in the haemostatic mechanism due to a physical interaction between volaemic colloids and individual clotting factors or formed elements in blood, and those alterations induced by haemodilution. It should also be emphasized that haemostasis is not synonymous with coagulation. In perhaps one-third of all patients with generalized bleeding, for example, no clotting abnormality is present (Alexander *et al.* 1975).

A vast amount of research has been performed describing the effect of hydroxyethyl starch on coagulation, fibrinolysis and haemostasis (Matsuda and Murakami 1972; Matsuda *et al.* 1972; Takiguchi 1977). In attempting to differentiate between changes brought about by simple haemodilution or those thought to be the result of complex formation, the following section will focus as much as the data will allow, on individual factors included in the various systems above, in order to identify those adversely affected by hydroxyethyl starch.

3.8.1. COAGULATION

3.8.1.1. INTRINSIC SYSTEM

Fibrinogen

Precipitates of several factors (fibrinogen, Factor VIII, and fibrin monomer) that participate in the intrinsic system, form at 0-8 °C, in plasma mixed with either dextran or hydroxyethyl starch (Alexander and Odake 1967; Alexander *et al.* 1975). The amount of fibrinogen precipitated, for example, is closely related to the fibrinogen:dextran ratio as well as to $M_{\bar{w}}$ and intrinsic viscosities of various dextran fractions. Hydroxyethyl starch is a less effective precipitant and *does not* exhibit this correlation.

The basic mechanism whereby dextran or hydroyethyl starch precipitates fibrinogen and Factor VIII is presumably related to physicochemical principles (i.e. molecular ionic radius and asymmetry of protein and/or colloid, solution, or water of hydration). However, despite repeated attempts, Alexander *et al.* (1975) were unable to demonstrate clear-cut complex formation between fibrinogen, dextran, or dydroxyethyl starch. Nevertheless, complexing could occur via non-covalent forces. Studies conducted at room temperature *in vitro* have shown that dilution of dog plasma in four parts of either dextran or HES 450/0.70 did not cause visible precipitation of fibrinogen (Lewis *et al.* 1966). Even though hydroxyethyl starch can precipitate some factors in the intrinsic system under conditions that would not be expected to occur during clinical use (e.g. infusion of colloid at 22-25 °C), measurement of fibrinogen titres *in vivo* may differentiate between changes induced by haemodilution alone, or by consumption and/or precipitate formation.

In rabbits (Irikura *et al.* 1972*d*) or dogs (Thompson and Gadsden 1965) bled 25-30 ml/kg and reinfused with 1.2-1.8 g/kg of either HES 40/0.55 or HES 450/0.70, respectively, the diminution in the concentration of fibrinogen in blood was attributed to haemodilution. A dilutional effect with a mild decrease in fibrinogen titres was also observed in monkeys given HES 150/0.70 alone or a frozen blood – HES 150/0.70 admixture to replace 25-37 per cent and 30-50 per cent of the calculated blood volume, respectively (Weatherbee *et al.* 1974*c*). Fibrinogen titres returned to normal levels 20-24 hours after injection.

In dogs given over 1.2 g/kg HES 450/0.70, however, the levels of fibrinogen decreased more than would be expected from haemodilution (Cheng *et al.* 1966; Garzon *et al.* 1967*a*; Karlson *et al.* 1967; Matsuura *et al.* 1973). Matsuura and his colleagues believed that the diminished fibrinogen titre was due to consumption; however, this was not substantiated.

In normal volunteers or patients given 15 (Janes *et al.* 1977), 30 (Inoue *et al.* 1977; Janes *et al.* 1977; Lee *et al.* 1968; Solanke 1968*a,b*), 45 (Janes *et al.* 1977), or 60 g (Ballinger *et al.* 1966*a*) of either HES 40/0.55 or HES 450/0.70 as a *single* infusion or 96–240 g HES 450/0.70 over several days (Lee *et al.* 1968; Martin *et al.* 1976; Peter *et al.* 1975; Vinazzer and Bergmann 1975; Watzek *et al.* 1978), changes in the concentration of fibrinogen were attributed to those produced by haemodilution. In patients given 30 g HES 450/0.70 immediately after the initiation of surgery and 30 g 2 hours later, hydoxyethyl starch was believed to interact with fibrinogen, resulting in the formation of precipitating complexes, even though demonstration of complex formation was not obtained (Müller *et al.* 1976, 1977; Popov-Cenić *et al.* 1976, 1977).

Recalcification time

Another method to determine the effect of hydroxyethyl starch on fibrinogen and Factor VIII is the use of the recalcification time and its modifications [partial thromboplastin time (PTT), activated PTT, and extended PTT]. These tests measure coagulant activity generated in the intrinsic system and are *prolonged* in deficiencies of fibrinogen and/or Factor VIII, as well as deficiencies of one or more of Factors XII, XI, X, IX, V, and prothrombin.

The plasma recalcification time or PTT is slightly prolonged in dogs (Matsuura *et al.* 1973), monkeys (Weatherbee *et al.* 1974*c*), and rabbits (Irikura *et al.* 1972*d*) given large quantities (0.6–3 g/kg) of either HES 40/0.55 or HES 150/0.70.

In normal subjects or patients given up to 60 g HES 450/0.70 as a *single* bolus injection, the PTT was slightly prolonged in several studies (Ballinger *et al.* 1966*a*; Martin *et al.* 1976) or remained unchanged or slightly shortened in others (Inoue *et al.* 1977; Mishler 1975; Mishler *et al.* 1976; Watzek *et al.* 1978; Zaffiri *et al.* 1969).

The PTT is slightly prolonged (Harke *et al.* 1976; Müller *et al.* 1976; Peter *et al.* 1975; Popov-Cenić *et al.* 1977; Rock and Wise 1979) or slightly shortened (Vinazzer and Bergmann 1975) in patients given a total of 60–120 g HES 450/0.70 over several days. In patients administered a total of 120–240 g (7.5 ml/kg per 12 hours) HES 450/0.70, the PTT was markedly shortened up to four days after injection, returning to normal by day seven (Lee *et al.* 1968).

Thrombin time

The thrombin time (TT) is also used as a rapid method for measuring fibrinogen concentration and detecting the presence of heparin or fibrinogen-split products. A prolonged time indicates a qualitative change in fibrinogen (titre < 100 mg/dl) or the presence of an antithrombic substance. A shortened time, on the other hand, indicates a fibrinoplastic property, that is, an increased rate of fibrinogen clotting by thrombin. Experiments *in vitro* have shown HES 450/0.70 in high doses to be fibrinoplastic (Lewis *et al.* 1966).

The TT is slightly shortened in monkeys (Weatherbee *et al.* 1974*c*) and patients (Müller *et al.* 1976; Peter *et al.* 1975; Popov-Cenić *et al.* 1977; Vinazzer and

Bergmann 1975) given a total of 60-120 g HES 150/0.70 or HES 450/0.70 either as a single injection or administered over several days. In two normal subjects given 60 g HES 450/0.70 as a single bolus injection or patients administered 0.3-0.6 g/kg HES 450/0.70 during cardiopulmonary bypass surgery, however, the TT was slightly prolonged (Ballinger *et al.* 1966*a*; Mishler *et al.* 1975*b*).

Factors VIII and IX

Factors VIII and IX have been shown to participate in the intrinsic clotting system. The activity of Factor VIII has been reported to be unchanged in patients receiving a total of 120 g HES 450/0.70 over three days (Vinazzer and Bergman 1975). Müller *et al.* (1976) and Popov-Cenić *et al.* (1977) have shown, however, that the activity of both Factor VIII and IX (both determined by first-stage methods) increased in patients after the second of two 30 g injections of HES 450/0.70. Under these same conditions, however, the activity of both Factor VIII and IX decreased if measured by a two-stage method.

Factors II, V, and X

Factors II (prothrombin), V, and X are common to both the intrinsic and extrinsic clotting schemes. The activities of Factors II, V, and X were all reduced by approximately 80 per cent when dogs were exchanged transfused with HES 450/0.70 down to a haematocrit of 10 per cent (Lewis *et al.* 1966). The activity of Factor V was reduced by 50 per cent in dogs after replacement of 30 per cent of the shed blood volume with an equal volume of HES 450/0.70 (Gollub and Schaefer 1968).

The activity of Factor V is decreased in patients given a total of 60 g HES 450/0.70 on two occasions on the same day (Harke *et al.* 1976; Popov-Cenić *et al.* 1977) or a total of 120 g HES 450/0.70 over three days (Vinazzer and Bergmann 1975).

The activities of Factors II and X were slightly decreased after the second of two 30 g infusions of HES 450/0.70 on the same day (Harke *et al.* 1976; Popov-Cenić *et al.* 1977).

Whole blood coagulation time

The whole blood coagulation time test is a measure of the overall activity of the intrinsic system. Prolongation of the clotting time is due to *marked* coagulation factor deficiences. The clotting time is not prolonged in dogs infused with 0.6 g/kg HES 450/0.70 (Cheng *et al.* 1966; Karlson *et al.* 1967). In rabbits (Irikura *et al.* 1972*d*) or dogs (Cheng *et al.* 1966; Garzon *et al.* 1967*a*; Karlson *et al.* 1967) given 1.2-1.8 g/kg of either HES 40/0.55 or HES 450/0.70, the clotting time is prolonged but returns to basal levels 4-24 hours after injection. The clotting time remains prolonged in dogs, 24 hours after a 2.4 g/kg infusion of HES 450/0.70 (Garzon *et al.* 1967*a*). The clotting time is not prolonged in patients or normal subjects administered 15-30 g HES 450/0.70 (Gollub *et al.* 1969*b*; Mishler 1975; Solanke 1968*a*, *b*; Takeyoshi *et al.* 1971*a*, *b*; Zaffiri *et al.*

1969). The clotting time is unchanged or slightly prolonged in normal subjects or patients administered 60–90 g HES 450/0.70 (Amakata *et al.* 1972*b*,*c*; Ballinger *et al.* 1966*a*; Inoue *et al.* 1977; Lee *et al.* 1968; Mishler *et al.* 1975*b*). The clotting time is prolonged in patients four days after the infusion of 120–240 g HES 450/0.70, returning to normal limits after one week (Lee *et al.* 1968).

3.8.1.2. EXTRINSIC SYSTEM

Prothrombin time

The one-stage prothrombin time (PT) measures the coagulant activity of the extrinsic system, including fibrinogen, prothrombin, and Factors V, VII, and X. The PT is prolonged when one or more of the following events occur: Factor V, VII, X, or prothrombin is deficient; the fibrinogen titre is less than 100 mg/dl; fibrin or fibrinogen-split products are present. The PT is not significantly altered *in vitro* when dog blood is diluted 75 per cent with HES 40/0.55 (Matsuura *et al.* 1973). The PT is slightly prolonged in monkeys given HES 150/0.70 to replace 25–33 per cent of their blood volume (Weatherbee *et al.* 1974*c*) or dogs given up to 3 g/kg HES 40/0.55 (Matsuura *et al.* 1973). The activity of prothrombin or Factor VII is reduced when dogs are either exchange-transfused with HES 450/0.70 to a haematocrit of 10 per cent (Lewis *et al.* 1966) or infused with 0.6–2.4 g/kg HES 450/0.70 (Garzon *et al.* 1967*a*; Karlson *et al.* 1967). The activity of prothrombin and Factor VII returned to normal limits in both of the above studies within 48 hours. The PT and the prothrombin consumption test were both markedly prolonged in rabbits up to 15 days after receiving 1.8 g/kg HES 40/0.55 (Irikura *et al.* 1972*d*). Prothrombin consumption is not, however, altered in monkeys given HES 150/0.70 to replace 25–33 per cent of the blood volume (Weatherbee *et al.* 1974*c*).

In normal subjects or patients administered single injections of 30–60 g HES 450/0.70, the PT or residual PT was either unchanged (Zaffiri *et al.* 1969), slightly shortened (Inoue *et al.* 1977; Mishler 1975; Mishler *et al.* 1976), or slightly prolonged (Ballinger *et al.* 1966*a*; Mishler *et al.* 1975*b*). In patients or normal subjects given single or multiple injections of HES 450/0.70 (total 60–120 g) the PT was shortened (Martin *et al.* 1976; Peter *et al.* 1975; Vinazzer and Bergmann 1975; Watzek *et al.* 1978) or remained unchanged (Rock and Wise 1979). Factor VII activity was decreased after two 30 g infusions of HES 450/0.70 on the same day (Popov-Cenić *et al.* 1977).

Reptilase time

An additional test often used to assess the extrinsic system is the reptilase time (RT). The RT is shortened in patients receiving two 30 g infusions of HES 450/0.70 on the same day (Müller *et al.* 1976; Popov-Cenić *et al.* 1977).

Feedback mechanisms

The inhibition of certain serine proteolytic enzyme factors used in both the intrinsic and extrinsic clotting systems serves as a negative feedback scheme.

In this regard, antithrombin III, α_1-antitrypsin and α_2-macroglobulin serve as inhibitors of serine proteolytic enzymes. In patients given two 30 g injections of HES 450/0.70 on the same day, antithrombin III and α_1-antitrypsin activities initially decreased, but returned to basal levels within 24 hours after injection (Müller *et al.* 1976. 1977; Popov-Cenić *et al.* 1976. 1977). Twenty-four hours after injection, levels of α_2-macroglobulin were approximately 80 per cent of the preinfusion value.

3.8.2. FIBRINOLYSIS

Fibrin deposition both intravascularly and extravascularly often occurs in health and disease. The resolution of such deposits is achieved *in vivo* through a basic repair mechanism including the enzymatic dissolution of insoluble fibrin polymers, a phenomenon referred to as fibrinolysis. Fibrinolysis is controlled and regulated chiefly by the activity of a normally circulating plasma proteolytic enzyme system termed the plasminogen–plasmin system. The activated enzyme plasmin is capable of digesting fibrin into a number of soluble fragments. The euglobulin lysis time (ELT) is a test used to evaluate systemic fibrinolysis. The ELT test is, however, not specific and is believed primarily to reflect activator activity.

The ELT is markedly reduced in rabbits infused with 1.8 g/kg HES 40/0.55. In these same animals, the fibrin heat plate time was also diminished (Irikura *et al.* 1972*d*). No alteration in fibrinolysis was noted in dogs given 0.6–2.4 g/kg HES 450/0.70 (Cheng *et al.* 1966). Monkeys given HES 150/0.70 to replace 25–33 per cent of their blood volume had no euglobulin lysis (Weatherbee *et al.* 1974*c*).

In patients, the ELT is diminished after a single 30 g infusion of HES 450/0.70 but returns to normal limits after a second 30 g injection on the same day (Müller *et al.* 1976, 1977; Popov-Cenić *et al.* 1976, 1977). In these same studies, the titre of plasminogen was reduced after the second of two 30 g infusions. The ELT was not significantly altered in patients administered either 30–90 g or 120–240 g HES 450/0.70 (Lee *et al.* 1968).

3.8.3. HAEMOSTASIS

Haemostasis is here defined as the overall arrest of haemorrhage.

3.8.3.1. BLEEDING TIME

The bleeding time (BT) is used as a screening test for haemostasis, as well as for disorders of platelet function.

The BT is progressively increased in dogs or hamsters in direct relationship to either the amount administered (Berman *et al.* 1965; Cheng *et al.* 1966; Garzon *et al.* 1967*a*; Gollub 1965; Gollub *et al.* 1967; Gollub and Schaefer 1968; Karlson *et al.* 1967; Schaefer *et al.* 1966) or the $M_{\overline{W}}$ of hydroxyethyl starch

(Gollub *et al.* 1969*a*). The BT in dogs is not significantly altered in the following circumstances: 0.6 g/kg HES 450/0.70 is injected or 10 per cent of the shed blood volume is replaced with HES 450/0.70 (Cheng *et al.* 1966; Garzon *et al.* 1967*a*; Gollub and Schaefer 1968; Gollub *et al.* 1967).

The BT is *significantly prolonged* in dogs in the following circumstances: infusion of 1.2–2.4 g/kg HES 450/0.70 (Cheng *et al.* 1966; Garzon *et al.* 1967*a*; Thompson and Gadsden 1965); replacement of 20–40 per cent of the shed blood volume by HES 450/0.70 (Gollub and Schaefer 1968; Gollub *et al.* 1967, 1969*a*; Schaefer *et al.* 1966); exchange replacement is conducted to a haematocrit of 10 per cent (Lewis *et al.* 1966). Under these conditions, the BT returns to normal limits within 10–48 hours (Thompson and Gadsden 1965; Lewis *et al.* 1966). Even under the most extreme of these conditions, hydroxyethyl starch produces a *milder* effect on the BT than either dextran 40 or dextran 60/75 (Gollub and Schaefer 1968; Gollub *et al.* 1967, 1969*a*; Karlson *et al.* 1967). The BT appears to be prolonged in direct relation to the $\bar{M}_w$ of hydroxyethyl starch: 310 000–1 490 000, producing greater effects than fractions with $\bar{M}_w$ of 40 000 (Gollub *et al.* 1969*a*).

In normal subjects or patients administered 30 g HES 450/0.70, the BT was not altered (Solanke 1968*a*, *b*; Maguire *et al.* 1980*a*; Takeyoshi *et al.* 1971*a*, *b*). The BT was *slightly prolonged*, however, in normal subjects or patients given 60 g or more of HES 450/0.70 (Amakata *et al.* 1972*b*,*c*; Ballinger *et al* 1966*a*). In patients receiving 0.3–0.6 g/kg HES 450/0.70 during cardiopulmonary bypass surgery, the BT was slightly prolonged 24 hours after surgery, but returned to normal limits after 48 hours (Mishler *et al.* 1975*b*).

3.8.3.2. BLOOD LOSS

The weighted blood loss and incidence of incisional rebleeding from standard skin incisions after colloid infusion are useful techniques to evaluate haemostasis. In the dog, blood loss or incidence of rebleeding is progressively increased in direct relationship to the quantity of HES 450/0.70 or dextran administered (Fig. 3.4). Blood loss or incisional rebleeding was not appreciable after infusion of 0.6 g/kg HES 450/0.70 (Cheng *et al.* 1966; Garzon *et al.* 1967*a*; Karlson *et al.* 1967) or replacement of 10 per cent of the shed blood volume (Gollub *et al.* 1967). Blood loss or incisional rebleeding is significiantly increased if 1.2–2.4 g/kg HES 450/0.70 is infused (Cheng *et al.* 1966; Garzon *et al.* 1967*a*; Karlson *et al.* 1967) or 20–40 per cent of the shed blood volume is replaced with HES 450/0.70 (Gollub *et al.* 1967). However, HES 450/0.70 has a *milder* effect on both blood loss (Karlson *et al.* 1967) and incisonal rebleeding (Gollub *et al.* 1967) when compared with either dextran 40 or dextran 60/75 (Fig. 3.4). No bleeding tendency or wound oozing was observed in monkeys given HES 150/0.70 to replace 25–33 per cent of their blood volume (Weatherbee *et al.* 1974*c*).

In patients undergoing surgery or cardiopulmonary bypass operations, blood loss was not increased over that expected from the severity of the procedure (Amakata *et al.* 1971; Hayashi and Higashi 1975 . Mishler *et al.* 1975*b*). In 26 patients given 15–30 g HES 450/0.70, only one experienced a minor temporary bleeding defect after extracorporeal circulation (Gollub *et al.* 1969*b*). In 29 patients receiving 30 g HES 450/0.70, two had bleeding but only 2 and 4 days after surgery (Solanke 1968*a*, *b*). Of 30 patients administered 30–240 g HES 450/0.70, one bled after heparinisation and two had minor bleeding from stress ulcers (Lee *et al.* 1968).

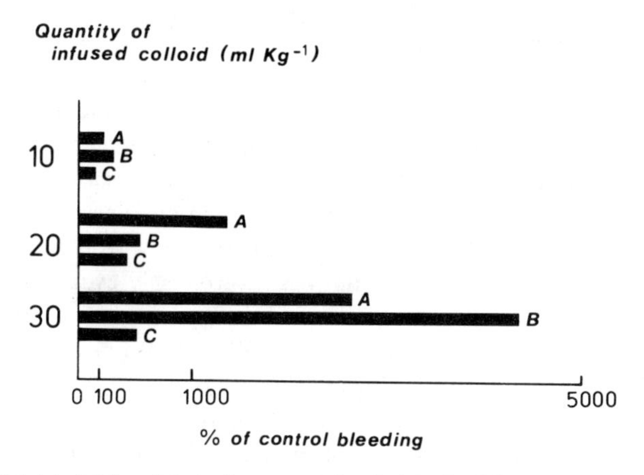

Fig. 3.4. Weighted blood loss from standard skin incisions in dogs after colloid (10, 20, and 30 ml/kg) infusion, expressed as the average per cent of control in each group. A = dextran 75; B = dextran 40; and C = HES 450/0.70. (From Karlson *et al.* 1967.)

3.8.3.3. THROMBOELASTOGRAPHY

The thromboelastogram (TEG) is a useful documentation of several properties of clot formation. The TEG is composed of four values: the value of *k* reflects the rate of build-up of the coagulum; the value of *ma* measures final tensile strength of the clot; the value of *me* indicates maximum elasticity; and the *r* value is the time for initial effective fibrin formation.

When blood is diluted *in vitro* with either HES 40/0.55 (Matsumoto *et al.* 1977) or HES 200/0.60 (Matsumoto *et al.* 1974), the TEG is not noticeably altered when one part hydroxyethyl starch is mixed with nine parts whole blood (10 per cent). When two parts hydroxyethyl starch are mixed with eight parts whole blood (20 per cent), the *r* and *k* values are slightly increased, whilst the value of *ma* is reduced. When a 30–75 per cent dilution of whole blood is achieved, the *k* and *r* values are significantly increased concomitantly with markedly reduced values of both *ma* and *me* (Matsumoto *et al.* 1974, 1977; Matsuura *et al.* 1973). Under these same test conditions, hydroxyethyl starch had a less pronounced effect on the TEG when directly compared with either dextran 40 or dextran 70.

In rabbits infused with 3.6 g/kg HES 40/0.55, the values of *r* and *ma* were significantly reduced four hours after injection, returning to normal limits after 24 hours (Irikura *et al.* 1972*d*). In rabbits given 1.8 g/kg per day of HES 40/0.55 for four days, the value of *r* was significantly reduced, whilst the value of *ma* was significantly prolonged after infusion. In these acutely and chronically infused rabbits, hydroxyethyl starch was shown to have a less pronounced effect on the TEG than dextran 70 under the same test conditions.

In dogs, after the replacement of more than 20 per cent of the shed blood volume with an equal volume of either HES 40/0.55 or HES 450/0.70, abnormalities in the *k* value as well as impaired *ma* and *me* values were noted (Gollub and Schaefer 1968; Matsuura *et al.* 1973). The *r* value, however, was nearly normal. Using the technique of nephelometry, Gollub and Schaefer (1968) noted an unexpected decrease in the absolute numbers (degree of light-scattering) of plasma clots after infusion of HES 450/0.70. The increased optical density was associated only with the process of clot formation. These investigators also noted that histological sections of plasma clots prepared from samples of blood drawn after HES 450/0.70 infusion in dogs showed only evenly dispersed, fine fibrin fibrils without the gross fasciculi of fibrin characteristic of normal plasma clots. Clots obtained from mixtures of HES 450/0.70 and plasma (>20 per cent of shed blood volume replaced with hydroxyethyl starch), both *in vivo* and *in vitro*, appeared friable. These clots were insoluble in 5 mol/l urea solutions. Clot retraction, however, was normal after large infusions, and this observation has been supported by additional studies in dogs (Cheng *et al.* 1966; Garzon *et al.* 1967*a*; Karlson *et al.* 1967; Lewis *et al.* 1966) and man (Balinger *et al.* 1966*a*; Inoue *et al.* 1977).

In patients administered 15–60 g HES 450/0.70, the *r* and *k* values were essentially unchanged, concomitant with slight reductions in the values of *ma* and *me* (Gollub *et al.* 1969*a*; Inoue *et al.* 1977; Müller *et al.* 1976; Popov-Cenić *et al.* 1977). In patients given 30 g HES 450/0.70 during surgery and 30 g on each of three following days (total 120 g), no significant alterations were observed for *r*, *k*, and *ma* values (Vinazzer and Bergmann 1975). In patients undergoing cardiopulmonary bypass surgery in which either 15–30 g (Gollub *et al.* 1969*b*) or 0.3–0.6 g/kg HES 450/0.70 (Mishler *et al.* 1975*b*) was administered, the values of *k*, *r*, and *ma* were essentially unchanged.

3.8.3.4. PLATELET COUNT AND FUNCTION

The initial aspects of haemostasis, as well as a normal bleeding time, are related both to the blood concentrations of platelets and to their function.

Platelet count

The concentration of platelets in mixtures of HES 40/0.55 and whole blood prepared *in vitro* are proportionally decreased as the amount of hydroxyethyl starch is increased (Matsuura *et al.* 1973). This relationship is also true in dogs given 0.6–3 g/kg of either HES 40/0.55 or HES 450/0.70 (Cheng *et al.* 1966; Garzon *et al.* 1967*a*; Karlson *et al.* 1967; Matsuura *et al.* 1973) or exchange-transfused with HES 450/0.70 to a haematocrit of 10 per cent (Lewis *et al.* 1966). In extreme circumstances where dogs are infused with up to 3 g/kg HES 450/0.70 or exchange-transfused to a haematocrit of 10 per cent, recovery of the platelet count to normal levels is obtained 4–24 hours in the former case, and eight days in the latter (Cheng *et al.* 1966; Garzon *et al.* 1967*a*; Karlson *et al.* 1967; Lewis *et al.* 1966). In monkeys given large quantities of a HES 150/0.70 – previously frozen blood admixture, the platelet count returns to near normal levels after 24 hours (Weatherbee *et al.* 1974*c*).

In normal subjects or patients administered 15–90 g of either HES 40/0.55 or HES 450/0.70, the concentration of platelets in whole blood decreases in direct proportion to the amount of hydroxyethyl starch infused (Amakata *et al.* 1971*a*,*c*; Ballinger *et al.* 1966*a*; Gollub *et al.* 1969*b*; Inoue *et al.* 1977; Maguire *et al.* 1979*b*; Mishler 1975; Peter *et al.* 1975; Solanke 1968*a*, *b*; Watzek *et al.* 1978; Zaffiri *et al.* 1969). Under the most extreme of these conditions, the platelet count returns to basal levels 24–48 hours after injection. In patients undergoing cardiopulmonary bypass surgery in which some degree of platelet destruction would be expected, the platelet count was decreased when either 15–30 g or 0.6–1.2 g/kg HES 450/0.70 was administered (Gollub *et al.* 1969*b*; Mishler *et al.* 1975*b*). The concentration of platelets rose to near normal values after seven days in patients receiving 0.6–1.2 g/kg (Mishler *et al.* 1975*b*).

Platelet function

The adhesion of platelets to the surface of an injured blood vessel constitutes one of the earliest stages in haemostasis. Platelet adhesion also plays a significant role in the formation of a thrombus. In dogs exchange-transfused with HES 450/0.70 down to a haematocrit of 10 per cent, platelet adhesiveness is markedly decreased concomitantly with a prolonged bleeding time (Lewis *et al.* 1966). Platelet adhesiveness is increased and the bleeding time is normal after 48 hours, and normal adhesion is observed after eight days. In these studies, HES 450/0.70 had a lesser effect on adhesion than either dextran 40 or dextran 75. In normal subjects given 30 g HES 450/0.70, platelet adhesiveness was not altered (Maguire *et al.* 1980*a*).

Phospholipid (platelet factor-3) is required at two stages in the formation of intrinsic prothrombinase and is normally derived from platelets. Platelet factor-4 is a glycoprotein released from platelets after platelet aggregation is induced by adenosine diphosphate (ADP), thrombin, or adrenalin. It shortens the thrombin clotting time in the presence of heparin, potentiates ADP-induced aggregation *in vitro*, precipitates fibrinogen, non-enzymatically clots soluble fibrin monomer complexes, and neutralizes certain fibrinogen breakdown products (antithrombin VI).

In patients given either two 30 g injections of HES 450/0.70 on the same day (total 60 g) or 120 g HES 450/0.70 over three days the levels of platelet factor-3 (either in platelet-rich or platelet-poor plasma) were unchanged (Müller *et al.* 1976, 1977; Popov-Cenić *et al.* 1976, 1977; Vinazzer and Bergmann 1975). In those patients administered two 30 g injections of HES 450/0.70 on the same day, the concentrations of platelet factor-4 (either in platelet-rich or platelet-poor plasma) were slightly decreased after the second infusion (Müller *et al.* 1976, 1977. Popov-Cenić *et al.* 1976, 1977).

Platelets participate in primary haemostasis by forming aggregates at the site of injured blood vessels. The agent which is ultimately responsible for platelet aggregation is probably ADP, which may be derived from injured tissue and erythrocytes or released from platelets themselves following reaction with, among others, collagen, thrombin, and adrenalin. An impairment of platelet aggregation will often manifest itself in a prolonged bleeding time.

Platelet aggregation is normal in platelets collected by continuous- or intermittent-flow centrifugation in which 15–45 g HES 450/0.70 are administered during the procedure to normal subjects (see Chapter 5 for details). In normal subjects receiving 30 g HES 450/0.70, spontaneous as well as collagen-, ADP-, and epinephrine-induced platelet aggregation were normal (Maguire *et al.* 1980*a*). In patients undergoing surgical procedures in which two 30 g injections of HES 450/0.70 were administered on the same day, the ability of platelets to aggregate under the stimulus of either collagen or adrenalin was decreased (Harke *et al.* 1976). Recovery of the ability to aggregate was observed after 28 hours. In these studies, similar decreases in platelet aggregation were observed in patients receiving infusions of either electrolytes or plasma proteins. In patients given 30 g HES 450/0.70 on the day of the surgical procedure and 30 g on each of three consecutive days (total 120 g), collagen- or adrenalin-induced platelet aggregation was significantly decreased (Vinazzer and Bergmann 1975). Patients receiving normal saline under the same experimental conditions as those patients given HES 450/0.70 experienced the same degree of reduced platelet aggregation.

3.8.4. SUMMARY

Whether hydroxyethyl starch elicits a *specific* effect on mechanisms of coagulation, fibrinolysis, or haemostasis, other than that of simple dilution remains somewhat unclear. Under extreme conditions (0–8 °C), hydroxyethyl starch appears to precipitate fibrinogen and Factor VIII, even though not as efficiently as dextran. In the region of 20–25 °C, however, precipitate formation is not observed. What is clear, however, is that hydroxyethyl starch will decrease the concentration of coagulation factors in direct proportion to the amount of colloid infused. In this context, it might be asked at what level of dilution of blood will untoward effects be observed? In man, the dilution of blood produced by the *single* infusion of 90 g (1500 ml) of hydroxyethyl starch may be considered the upper limit of safety. This level is taken from the studies in dogs, in which the tensile strength and fibrin organization in clots is altered when more than 20 mg/ml hydroxyethyl starch are present in plasma. A plasma concentration of 20 mg/ml in man is normally produced by the infusion of 1500 ml of hydroxyethyl starch.

3.9. EFFECTS ON ORGAN FUNCTION

3.9.1. LIVER

3.9.1.1. RABBITS

No significant changes were observed in serum levels of aspartate aminotransferase (AST), alanine aminotransferase (ALT), and bilirubin in rabbits treated intravenously with the following dosage schedules: 3 g/kg per day HES 140/0.65 for one or three weeks (Lindblad 1970); 1.8 or 3.6 g/kg per day HES 40/0.55 for thirty days (Irikura *et al.* 1972*g*); 0.9, 1.8, or 3.6 g/kg per day HES 40/0.55 for three months (Irikura 1972); and 0.7, 2.1, or 6.3 g/kg per day HES 450/0.55 for thirty days (Irikura *et al.* 1972*l*). Concentrations of cholesterol, triglycerides, and phospholipids in liver were not affected in animals given 1.8 g/kg per day of HES 40/0.55 for four days (Irikura *et al.* 1972*q*). In rabbits receiving HES 40/0.55, no alterations in blood activities of ALT or AST were noted (Koyama and Yamauchi 1972).

3.9.1.2. RATS

In rats administered either 60, 120, or 240 mg/kg per day of HES 450/0.70 for twenty-eight days, blood concentrations of AST were normal whilst increases in ALT activity were noted (Odaka *et al.* 1971*a*). In rats given either 2.4 or 4.8 g/kg per day of HES 450/0.70 for five days, levels of AST and ALT were normal (Odaka *et al.* 1971*b*).

3.9.1.3. DOGS

In dogs receiving 3 g/kg per day of HES 140/0.65 for three weeks, blood ALT activities were normal, concomitantly with increased amounts of AST (Lindblad 1970).

3.9.1.4. MAN

In normal subjects or patients injected intravenously with up to 60 g of HES 150/0.70 (Mishler *et al.* 1978*a*), HES 200/0.60 (Yamamoto *et al.* 1974), HES 264/0.43 (Mishler *et al.* 1979*b*; Köhler *et al.* 1977*e*), HES 350/0.60 (Mishler and Dürr 1980) or HES 450/0.70 (Köhler *et al.* 1977*e*; Mishler 1975; Mishler *et al.* 1976; Takeyoshi *et al.* 1971*a*, *b*), blood levels of AST, ALT, total bilirubin, lactic dehydrogenase (LDH), alkaline phosphatase (AP), and gamma-glutamyl transferase were all normal. In normal subjects given a single 30 g injection of HES 450/0.70 on three or four consecutive days (total 90–120 g) (Mishler 1978*a*; Rock and Wise 1979) or over 96 hours (total 90 g) (Mishler *et al.* 1974*a*), blood levels of bilirubin, AST, ALT, and LDH were normal after the infusion period. Normal subjects undergoing leucapheresis on four or more occasions had normal hepatic-screening test results when given HES 450/0.70 during each procedure (see Chapter 5 for details).

In patients receiving 30–60 g of either HES 40/0.55 or HES 450/0.70 during various surgical procedures, blood levels of total bilirubin, acid phosphatase, beta-glucuronidase, AST, ALT, AP, and LDH were usually within normal limits after the infusion period, even although in a small number of cases LDH or ALT activities were transiently increased (Adachi *et al.* 1972; Amakata *et al.* 1972*b*, *c*; Goto *et al.* 1972*b*; Kobayashi *et al.* 1971; Kudo *et al.* 1971; Lamke and Liljedahl 1977; Mishler *et al.* 1975; Nakajo 1972; Okada *et al.* 1971; Oishi 1971; Yoshitake *et al.* 1971). These increases in LDH and ALT activities were, however, not pronounced.

3.9.2. KIDNEY

3.9.2.1. RABBITS

In dehydrated rabbits (Yamasaki 1975) given 3.6 g/kg HES 40/0.55, the blood urea nitrogen concentration (BUN) and the clearance of creatinine (C_{cr}) were not altered after infusion.

Rabbits have been treated intravenously according to the following schedules: 1.8 or 3.6 g/kg per day HES 40/0.55 for thirty days (Irikura *et al.* 1972*g*); 0.7, 2.1, or 6.3 g/kg per day HES 450/0.55 for thirty days (Irikura *et al.* 1972*l*); and 0.9, 1.8, or 3.6 g/kg per day HES 40/0.55 for three months (Irikura 1972). Abnormal quantities of occult blood, glucose, bilirubin, and ketones were not detected in voided urine. Slight amounts of protein and urobilinogen were, however, found in urine. The amount and specific gravity of voided urine were not altered in comparison with control animals (Irikura *et al.* 1972*g*). In rabbits given either 0.6 or 1.8 g/kg per day HES 40/0.55 for ten days, renal blood flow (RBF) and glomerular filtration rate (GFR) were not significantly changed, although a slight decrease was observed in the filtration fraction (FF) (Irikura *et al.* 1972*b*). A significant decrease in GFR and FF were observed, however, in animals treated with 5.4 g/kg per day for ten days with either HES 40/0.55 or dextran 70 (Irikura *et al.* 1972*b*). However, a significant increase of RBF was noted in animals treated with HES 40/0.55.

3.9.2.2. DOGS

The flow of urine, as well as the clearance of both creatinine and para-amino-hippuric acid (PAH), were increased in normotensive splenectomized and intact dogs after the infusion of 1.2 g/kg HES 450/0.70 (Murphy *et al.* 1965). In dogs made hypotensive and then resuscitated with 1.2 g/kg HES 450/0.70, the flow of urine and clearance of PAH were increased over control levels 24 hours after infusion (Murphy *et al.* 1965). In these circumstances, the C_{cr} was slightly decreased. Urinary output was normal in dogs exchange-transfused with HES 450/0.70 down to a haematocrit of 10 per cent (Takaori *et al.* 1968). In dogs subjected to acute haemorrhagic shock in which the total quantity of blood withdrawn was replaced with HES 450/0.70, GFR and the C_{cr} were returned to preshock levels after infusion (Heidenreich *et al.* 1975).

In dogs given either 3 g/kg per day HES 140/0.65 for three weeks (Lindblad 1970) or 7.8–15 g/kg HES 450/0.70 in 2–5 days (Thompson 1963; Thompson and Walton 1962, 1963), C_{cr}, glucose reabsorption, PAH excretion, glucose concentration, and urine specific gravity and pH were all within normal limits, even although, in the study of Lindblad, small amounts of protein appeared irregularly in the urine of all dogs.

3.9.2.3. MONKEYS

In monkeys that had 30-50 per cent of their calculated blood volume replaced with either autologous or homologous blood previously frozen in a 14 per cent mixture of HES 150/0.70, the blood concentration of creatinine was normal up to 98 hours after transfusion, while BUN was slightly increased during this period (Weatherbee *et al.* 1974*c*). The increases were not pronounced and were temporary.

3.9.2.4. MAN

In normal subjects given a single injection of up to 60 g of HES 150/0.70, HES 200/0.60, HES 264/0.43, HES 350/0.60, or HES 450/0.70, urinary output, C_{cr}, and urinary creatinine concentration, as well as blood concentrations of creatinine, urea, and uric acid, were all within normal limits after infusion (Mishler *et al.* 1978*a*, 1979*b*; Mishler and Dürr 1980; Yamamoto *et al.* 1974).

The results of renal screening tests in subjects receiving HES 450/0.70 during leucapheresis are presented in Chapter 5 (p. 129).

Blood creatinine, uric acid, and BUN concentrations were also normal in subjects given a 30 g infusion of HES 450/0.70 on three or four consecutive days (total 90 or 120 g) (Mishler *et al.* 1977*b*; Rock and Wise 1979). In patients undergoing various surgical procedures in which up to 90 g of HES 40/0.55, HES 200/0.60, or HES 450/0.70 were administered, C_{cr} and urinary creatinine as well as blood BUN and creatinine concentrations were within normal limits (Adachi *et al.* 1972; Hempel *et al.* 1975; Lee *et al.* 1968; Goto *et al.* 1972*b*; Kobayashi *et al.* 1971; Kudo *et al.* 1971; Lamke and Liljedahl 1977; Okada *et al.* 1971; Nakajo 1972; Takeyoshi *et al.* 1971*a*, *b*; Yamamoto and Momose 1972; Yamamoto *et al.* 1974; Watzek *et al.* 1978). In some patients receiving HES 450/0.70, the volume of urine voided was below that expected (Hayashi and Higashi 1975; Kobayashi *et al.* 1971), even though in other patients given HES 40/0.55, HES 200/0.60, or HES 450/0.70, urine volumes were normal (Adachi *et al.* 1972; Amakata *et al.* 1972*b*; Goto *et al.* 1972*b*; Hempel *et al.* 1975; Kitamura *et al.* 1972*a*,*b*; Lazrove *et al.* 1980; Yamamoto *et al.* 1974; Watzek *et al.* 1978). The clearance of phenolsulphonephthalatein was decreased in 40 per cent of patients receiving HES 450/0.70 (Amakata *et al.* 1972*c*), even though normal values were observed in patients also given HES 450/0.70 (Lee *et al.* 1968) or HES 200/0.60 (Nakajo 1972).

The effect of hydroxyethyl starch during perfusion of kidneys *in vitro* will be discussed in Appendix 3.

3.9.3. MISCELLANEOUS

3.9.3.1. RABBITS

The glucose tolerance test, as well as blood concentrations of cholesterol, triglycerides, free fatty acids and phospholipids, were not altered in rabbits treated intravenously with 1.8 g/kg per day HES 40/0.55 for four consecutive days (Irikura *et al.* 1972*q*). Intravenous administration of 1.5 or 3 g/kg per day HES 40/0.55 for four days had no effect on the drug-metabolizing enzyme activity of aniline hydroxylase and aminopyrine N-demethylase (Irikura *et al.* 1972*q*).

3.9.3.2. MAN

The concentration of cholesterol in blood was not affected in normal subjects (Mishler *et al.* 1978*a*) or patients (Nakajo 1972) given up to 56 g of either HES 150/0.70 or HES 200/0.60. The blood insulin concentration and the ratio of plasma insulin to plasma glucose rose 30–60 minutes after patients received 44–54 g of HES 40/0.55 (Adachi *et al.* 1972). These increases subsided after 90 minutes. The concentration of lipase in blood was not affected in normal subjects or patients dosed with up to 60 g of either HES 350/0.60 (Mishler and Dürr, 1980) or HES 450/0.70 (Köhler *et al.* 1977*e*).

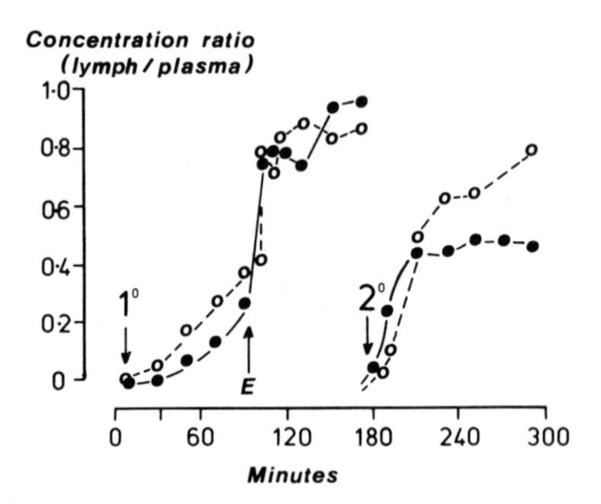

Fig. 3.5. The concentration ratios of thoracic duct lymph and plasma in dogs administered albumin (○) or HES 450/0.70 (●) with (1°) or without (2°) endotoxin (E). (From Chien *et al.* 1965.)

3.10. CONCENTRATION OF HYDROXYETHYL STARCH IN LYMPH AND THE RATE OF LYMPH FLOW AFTER ACUTE ADMINISTRATION

In the dog, the concentration of either HES 40/0.55 (Fukutome 1977) or HES 450/0.70 (Chien *et al.* 1965; Fukutome 1977) in lymph gradually increases after intravenous administration (Fig. 3.5), thus accounting for the PAS-positive granules found in lymph-node macrophages after dosing (see Section 2.3.2.2).

Even though HES 450/0.70 persists longer in lymph than does HES 40/0.55, the magnitude of the initial increase and subsequent duration of the colloidal osmotic pressure of lymph remains the same after infusion of either colloid (Fukutome 1977). In dogs challenged with endotoxin, greater amounts of HES 450/0.70 appear in lymph compared with non-endotoxin-treated animals (Fig. 3.5). This increase in the ratio of lymph to plasma after injection of endotoxin is due to true permeability changes in the hepatic and intestinal circulations (Chien *et al.* 1965).

Controversy exists, however, over whether hydroxyethyl starch induces a change in the rate of thoracic-duct lymph flow after administration. Fukutome (1977) claimed that the rate of thoracic-duct lymph flow increased in dogs treated intravenously with either HES 40/0.55 or HES 450/0.70, whereas Gollub and his colleagues (1967) reported that the rate of thoracic-duct lymph flow did not increase in dogs treated with HES 450/0.70.

4. Experimental and clinical use of hydroxyethyl starch as a volaemic colloid

4.1. INTRODUCTION

It is well known that restoration of blood volume in patients that have sustained a blood loss will improve blood perfusion of vital organs (Thompson 1974). In most patients with hypoperfusion who have no obvious blood loss or hypovolaemia, augmentation of blood volume and heart filling will increase cardiac output and vital organ perfusion. In circumstances where ventricular filling pressures are increased, pulmonary oedema will occur when pulmonary capillary hydrostatic pressures exceed effective plasma colloidal osmotic pressure. The choice of fluids for augmentation of blood volume is thus critical in all these cases. Saline, either as sodium chloride or mixtures with other salts, is appropriate for a very transient increase in the blood volume, because approximately 25 per cent of these solutions will remain intravascular and continued administration will decrease the concentration of plasma proteins in blood; concomitantly the colloidal osmotic pressure will be lowered, causing peripheral and pulmonary oedema. Hydroxyethyl starch has been shown, both in animal and clinical

TABLE 4.1 *Survival (%) of dogs after administration of various resuscitating fluids*

(a) 24-hour mortality

Resuscitating fluid	Percentage of shed blood (volume) 10	20	30	40	Reference
Whole blood	–	–	57	69	Gollub *et al.* (1967)
No replacement	–	33	50	–	
HES 450/0.70	0	17	33	50	
Ringer's lactate	17	0	0	67	
Dextran 70	0	0	0	33	
Dextran 40	2	17	0	33	

(b) 72-hour survival after haemorrhagic shock

Resuscitating fluid	Survival (%) after resuscitation following 75 minutes of haemorrhagic shock	Reference
Whole blood	72	Murray *et al.* (1965)
No replacement	20	Ballinger *et al.* (1966*b*)
Dextran 70	44	Ballinger (1965)
HES 450/0.70	64	Ballinger (1965)

models, to augment blood volume and, more importantly, to maintain colloidal osmotic pressure. This chapter will focus on the effects of hydroxyethyl starch on various indices of hypovolaemia and haemodilution. Hence the clinician will be better able to predict the efficacy of hydroxyethyl starch in the management of patients requiring augmentation of a deficient blood volume.

TABLE 4.2 *Survival of dogs after administration of various resuscitating fluids, after endotoxin shock**

Resuscitating fluid	Survival after endotoxin shock (%) lasting: 2 days	3 days
Control	9.1	9.1
Dextran	22	0
HES 150/0.70	40	0
HES 450/0.70	80	80

*From Evangelista *et al.* (1969), Brown *et al.* (1968)

4.2. SURVIVAL OF ANIMALS SUBJECTED TO EITHER HAEMORRHAGIC SHOCK OR SEVERE HAEMODILUTION

One criterion used to assess the efficacy of colloid replacement after initiation of haemorrhagic shock or severe haemodilution is the number of test animals surviving such treatment. When shock or haemodilution is produced by various techniques in various test models, hydroxyethyl starch has been found to be a useful resuscitating fluid. In most instances, hydroxyethyl starch has been shown to be as good as whole blood or plasma (Tables 4.1 and 4.6), and equal to (Tables 4.1, 4.3, 4.4, and 4.6) or better than (Tables 4.1, 4.2, 4.4, and 4.6) either dextran or Ringer's lactate.

TABLE 4.3 *Survival of rats after administration of hydroxyethyl starch or dextran, after haemodilution**

Resuscitating fluid	Survival (%) after haemodilution to 20% haematocrit
HES 450/0.70	100
Dextran 60	100

* From Hölscher (1973/74)

4.3. CHANGES IN BLOOD AND PLASMA CONSTITUENTS AFTER RESUSCITATION

4.3.1. HAEMATOCRIT, HAEMOGLOBIN, AND LEUCOCYTES

The extent of changes in the concentration of erythrocytes, haemoglobin, and leucocytes in blood together with alterations in the haematocrit after acute

TABLE 4.4 *Survival of rabbits after administration of various resuscitating fluids, after various protocols**

Resuscitating fluid	Survival (%) after protocol:			
	A	B	C	D
HES 40/0.55	100	100	85	100
Dextran 70	100	100	87	87
Dextran 40	100	–	87	100
Ringer's lactate	100	20	47	67

A = 20 ml/kg blood withdrawn and replaced with an equal volume of test solution
B = Protocol A repeated twice
C = 30 ml/kg blood withdrawn and replaced with an equal volume of test solution
D = 30 ml/kg blood withdrawn and replaced with 45 ml/kg test solution
* From Irikura *et al.* (1972*e*)

TABLE 4.5 *24-hour survival of dogs after administration of various resuscitating fluids in replacement exchange**

Resuscitating fluid	Survival (%)
Dextran 70	100
Gelatin	83
Hydroxyethyl starch[1]	100
Hydroxyethyl starch[2]	67
Hydroxyethyl starch[3]	83
Hydroxyethyl starch[4]	100
Hydroxyethyl starch[5]	100
Hydroxyethyl starch[6]	83

1 – $M_{\bar{W}}$ 70 000 in Ringer's lactate; 2 – $M_{\bar{W}}$ 450 000; 3 – $M_{\bar{W}}$ 40 000; 4 – $M_{\bar{W}}$ 120 000
5 – $M_{\bar{W}}$ 310 000; 6 – $M_{\bar{W}}$ 1 490 000
* Gollub *et al.* (1969*a*)

TABLE 4.6 *Survival and mortality of dogs after administration of various resuscitating fluids*

(a) Survival after severe haemodilution

Resuscitating fluid	Survival (%)	Reference
HES 450/0.70	100	Takaori *et al.* (1970)
Dextran 40	100	

(b) Mortality after haemorrhagic shock

Resuscitating fluid	Mortality (%)	Reference
Dextran 70	50	Vineyard *et al.* (1966)
Dextran 40	60	
Plasma	90	
Ringer's lactate	80	
HES 450/0.70	85	

administration of hydroxyethyl starch depends on the amount of colloid infused, the water-binding potential of the infused species (effects *in vitro* should be distinguished from action *in vivo*, where production of small, osmotically active molecules will increase, thus increasing the amount of non-active water bound with time after infusion), the intravascular persistence of the injected species, and whether whole blood or plasma was withdrawn before infusion of colloid.

4.3.1.1. ANIMALS

In most circumstances in animals, the blood concentration of erythrocytes, haemoglobin, and leucocytes, as well as the haematocrit, will be decreased in direct proportion to the amount of any species of hydroxyethyl starch administered (Dillon *et al.* 1966; Gollub *et al.* 1967; Hartung *et al.* 1979; Hölscher 1973/74; Ikeda *et al.* 1971; Irikura *et al.* 1972*g*; Jesch *et al.* 1975; Thompson and Walton 1963; Vineyard *et al.* 1966). This relationship occurs whether whole blood or plasma has been withdrawn before infusion.

In animal studies comparing the persistence of the haemodilutional effect, a direct correlation was observed between duration of haemodilution and the intravascular survival of the species of hydroxyethyl starch injected. The concentration of haemoglobin and leucocytes, as well as the haematocrit, returned to normal preinjection levels *sooner* after rapidly metabolized colloids, such as HES 40/0.55 or dextran 40, were injected, in comparison with a slower effect produced by the more slowly catabolized HES 450/0.70 (Fukutome 1977; Irikura *et al.* 1972*e*; Takaori *et al.* 1968, 1970).

The ability of some species of hydroxyethyl starch to bind greater amounts of intravascular water (see Section 4.3.7) would normally exert a greater influence on the degree of haemodilution achieved after acute infusion. Even though smaller molecular weight species (e.g. HES 40/0.55) do bind more water, they are metabolized rapidly and the net effect is that they do not exert a greater haemodilutional effect than larger molecular weight species that bind less water but persist longer (Fukutome 1977; Sudo *et al.* 1972). Smaller molecular weight species of hydroxyethyl starch appear to be extravasated more easily than larger species, and this effect would also tend to decrease their effect on haemodilution (see Section 4.6).

Even though the infusion of hydroxyethyl starch produces similar haemodilutional decreases in the blood concentrations of erythrocytes, haemoglobin, and leucocytes, the numbers of reticulocytes increase after haemodilution and remain above preinfusion levels, even after concentrations of the other mentioned constituents have returned to basal values (Hölscher 1973/74; Hölscher and Kagel 1976; Irikura et al. 1972*m*).

In circumstances where the infusion of hydroxyethyl starch is administered over weekly or monthly intervals, the decrease in the haematocrit and in the concentrations of erythrocytes, haemoglobin, and leucocytes is directly related to the amount and intravascular persistence of the species injected (Hölscher 1972; Lindblad 1970; Irikura 1972; Irikura *et al.* 1972*g*,*l*). In these circumstances, the leucocyte differential count is not significantly affected (Irikura 1972; Irikura *et al.* 1972*g*, *l*).

4.3.1.2. MAN

If a single injection of hydroxyethyl starch is administered to patients or normal subjects, decreases in the haematocrit and in blood concentrations of erythrocytes, haemoglobin, and leucocytes normally follow the relationships previously mentioned for animals: the haemodilutional effect is directly related to the amount of colloid given, and the duration of this effect is directly related to the persistence of the species of hydroxyethyl starch injected (Adachi *et al.* 1972; Boon *et al.* 1976; Ehrly *et al.* 1979; Goto *et al.* 1972*b*; Hayashi and Higashi 1975; Homann *et al.* 1977; Kitamura *et al.* 1972*a*,*b*; Kraatz *et al.* 1975; Lamke and Liljedahl 1977; Lee *et al.* 1968; Metcalf *et al.* 1970; Mishler *et al.* 1978*a*; Nakajo 1972).

4.3.2. TOTAL PROTEIN

4.3.2.1. ANIMALS

With various animal models that produce either acute haemorrhagic shock or severe haemodilution, the absolute concentration of total plasma protein is reduced in direct proportion to the amount of any species of hydroxyethyl starch injected (Dillon *et al.* 1966; Ikeda *et al.* 1971; Irikura *et al.* 1972*e*,*f*,*n*; Jesch *et al.* 1975; Takaori *et al.* 1968, 1970; Weatherbee *et al.* 1974*c*), even though in severe haemodilution the fall in total protein may be greater than the corresponding decrease in the haematocrit (Lewis *et al.* 1966). The average plasma protein electrophoretic distribution pattern after severe haemodilution may also be altered (Lewis *et al.* 1966). The absolute decrease in albumin after severe exchange-transfusion with hydroxyethyl starch probably indicates that albumin synthesis rates are much slower than those of globulins, because all globulin concentrations return to normal much sooner than does albumin concentration. The persistence of induced hypoproteinaemia is directly related to the intravascular survival of the species of hydroxyethyl starch injected (Irikura *et al.* 1972*n*; Takaori *et al.* 1968, 1970; Lewis *et al.* 1966).

In rabbits given large quantities of either HES 40/0.55 (0.9–3.6 g/kg) or HES 450/0.55 (0.6–5.4 g/kg) daily for thirty days, total plasma protein was reduced in direct proportion to the amount and intravascular survival of the species of hydroxyethyl starch administered (Irikura 1972; Irikura *et al.* 1972*g*, *l*). In these same animals, the absolute concentration of albumin and globulins, as well as the ratio of albumin to globulins, was not adversely affected if the dose of either HES 40/0.55 or HES 450/0.55 was under 0.9 g/kg. Moderate decreases in all of the above levels were observed if the amount of hydroxyethyl starch was over 1.8 g/kg, even though animals given rapidly metabolized HES 40/0.55 had a less pronounced reduction. In rabbits receiving 0.9–3.6 g HES 40/0.55/kg for three months, moderate reductions in the absolute concentration of albumin and globulins and in the ratio of albumin to globulins were observed when animals were given over 1.8 g/kg (Irikura 1972).

4.3.2.2. MAN

In patients or volunteers receiving various species of hydroxyethyl starch, three generalizations can be made. Firstly, the reduction of total plasma protein after injection is directly related to the volume and intravascular survival of the species of hydroxyethyl starch used. Secondly, the absolute concentration of albumin is reduced more than globulins. Thirdly, the decrease in plasma protein levels is normally slightly greater than the corresponding fall in the haematocrit (Adachi *et al.* 1972; Boon *et al.* 1976; Ehrly *et al.* 1979; Goto *et al.* 1972*b*; Jesch *et al.* 1979; Kobayashi *et al.* 1971; Lamke and Liljedahl 1977; Lee *et al.* 1968; Metcalf *et al.* 1970; Mishler *et al.* 1976, 1978*b*; Rock and Wise 1979; Yamamoto *et al.* 1974).

4.3.3. PLASMA ELECTROLYTES

4.3.3.1. SODIUM

The plasma sodium concentration did not change significantly in various animals receiving large quantities of any species of hydroxyethyl starch (Heidenreich *et al.* 1975; Ikeda *et al.* 1971; Irikura 1972; Irikura *et al.* 1972*f*, *l*; Lindblad 1970; Odaka *et al.* 1971*a*, *b*; Sudo *et al.* 1972; Takaori *et al.* 1970; Yamasaki 1975). This observation was true whether hydroxyethyl starch was administered as a single injection or infused daily for periods of up to 30 days, and whether hydroxyethyl starch was suspended in 0.9 per cent isotonic saline or Ringer's lactate.

In patients or normal subjects given a *single* injection with any species of hydroxyethyl starch, the plasma sodium concentration may be slightly reduced (1–7 mEq/l) after injection. The decrease in sodium is not pronounced and is temporary (Adachi *et al.* 1972; Ballinger *et al.* 1966*a*; Goto *et al.* 1972*b*; Kitamura 1972*a*,*b*; Kobayashi *et al.* 1971; Kudo *et al.* 1971; Lamke and Liljedahl 1977; Lee *et al.* 1968; Nakajo 1972; Takeyoshi *et al.* 1971*a*,*b*; Watzek *et al.* 1978; Yamamoto and Momose 1972). The reduction in plasma sodium was observed whether hydroxyethyl starch was suspended in 0.9 per cent isotonic saline or Ringer's lactate. In normal subjects administered 30 g HES 450/0.70 on four consecutive days (total 120 g), there was no significant alteration in the plasma sodium concentration (Rock and Wise 1979).

4.3.3.2. POTASSIUM

In most animal models in which large quantities of various species of hydroxyethyl starch have been administered either as a single injection or given over long periods of time, the plasma potassium concentration has remained essentially unchanged (Heidenreich *et al.* 1975; Irikura 1972; Irikura *et al.* 1972*l*; Jesch *et al.* 1975; Lindblad 1970; Odaka *et al.* 1971*a*, *b*; Sudo *et al.* 1972; Takaori *et al.*

1970; Weatherbee *et al.* 1974*c*). Temporary reductions or increases in plasma potassium concentrations have been observed in dehydrated rabbits (Yamasaki 1975) and bled dogs (Ikeda *et al.* 1971; Irikura *et al.* 1972*e*), respectively. These alterations, however, were not pronounced.

In patients or normal volunteers given a total of up to 120 g of any species of hydroxyethyl starch, the concentration of plasma potassium after single or multiple injections has remained essentially unchanged (Adachi *et al.* 1972; Ballinger *et al.* 1966*a*; Goto *et al.* 1972*b*; Kitamura *et al.* 1972*a,b*; Kobayashi *et al.* 1971; Kudo *et al.* 1971; Lamke and Liljedahl 1977; Lee *et al.* 1968; Nakajo 1972; Rock and Wise 1979; Takeyoshi *et al.* 1971*a,b*; Watzek *et al.* 1978; Yamamoto and Momose 1972).

4.3.3.3. CHLORIDE

The plasma chloride concentration in various animals administered different species of hydroxyethyl starch remained unchanged after injection (Ikeda *et al.* 1971; Irikura 1972; Irikura *et al.* 1972*b*; Odaka *et al.* 1971*a,b*; Sudo *et al.* 1972; Yamasaki 1975). This observation was true whether large quantities of hydroxyethyl starch were administered as a single infusion or injected over long periods of time. Plasma chloride concentrations remained essentially unchanged in patients or normal subjects administered a total of up to 120 g of any species of hydroxyethyl starch (Adachi *et al.* 1972; Ballinger *et al.* 1966*a*; Goto *et al.* 1972*b*; Kobayashi *et al.* 1971; Kudo *et al.* 1971; Lee *et al.* 1968; Nakajo 1972; Rock and Wise 1979; Takeyoshi *et al.* 1971*a*; Watzek *et al.* 1978; Yamamoto and Momose 1972).

4.3.3.4. CALCIUM

Dogs or rabbits given various species of hydroxyethyl starch either acutely or chronically for up to three months showed no significant changes in plasma calcium concentration (Irikura 1972; Irikura *et al.* 1972*f*, *l*; Lindblad 1970).

Slight and temporary reductions in the plasma calcium concentration were observed in patients administered either HES 40/0.55 (Goto *et al.* 1972*b*) or HES 450/0.70 (Kobayashi *et al.* 1971; Lamke and Liljedahl 1977; Takeyoshi *et al.* 1971*a*; Watzek *et al.* 1978). These alterations, however, were not pronounced and concentrations usually returned to normal within 24 hours after injection.

4.3.3.5. MAGNESIUM

In patients given HES 450/0.70, the plasma magnesium concentration after injection was essentially unaltered (Watzek *et al.* 1978).

4.3.4. INTERMEDIARY METABOLISM

Under normal aerobic conditions, pyruvate serves as a precursor to several intermediates necessary for the tricarboxylic acid (Kreb's) cycle. During anaerobic glycolysis, which normally results from oxygen debt, glycogen is broken down to pyruvate and then to lactic acid. Thus, a rise in the blood lactate concentration reflects some degree of tissue hypoxia. In various models producing haemorrhagic shock in which inadequate blood flow and tissue ischaemia persist, the liver is no longer able to resynthesize glycogen from lactate, and blood lactate concentration rises in direct relation to the intensity and duration of shock.

4.3.4.1. ANIMALS

In a variety of models producing severe haemorrhagic shock in dogs and cats, the blood lactic acid concentration is greatly increased at the end of the shock period but returns to basal levels after the animal has been resuscitated with either HES 40/0.55 or HES 450/0.70 (Ikeda *et al.* 1971; Irikura *et al.* 1972*f*; Takaori and Safar 1976; Takaori *et al.* 1968; Vineyard *et al.* 1966; Zimmermann 1971; Zimmermann and Bannert 1970). In these studies, the blood pyruvate concentrations were essentially unaltered (Ikeda *et al.* 1971; Takaori *et al.* 1968).

4.3.4.2. MAN

Blood lactic acid and pyruvate concentrations in patients that had sustained minor blood loss were normal 24 hours after the patients were resuscitated with either HES 40/0.55 (Adachi *et al.* 1972; Grünert *et al.* 1978) or HES 450/0.70 (Adachi *et al.* 1972; Lee *et al.* 1968).

4.3.5. BUFFER BASE AND PH

Buffers are substances that tend to stabilize the pH of a solution. They are partially ionized salts formed by the combination of a strong acid and a weak base, or *vice versa*. A buffer is effective only when there are appreciable amounts of the salt and either the acid or the base from which it is derived. It is most effective when the salt and the acid or the base are present in equal quantities. For this reason, buffers are usually thought of in terms of buffer pairs. Common buffer pairs in blood are $NaHCO_3/H_2CO_3$ and Na_2HPO_4/NaH_2PO_4. The proteins of the blood – haemoglobin, oxyhaemoglobin, albumin, and globulin – are extremely important buffers. Owing to the amphoteric character of proteins, each species represents a buffer pair. The most important protein buffer in blood is haemoglobin, because of its high concentration. It is considerably more important than the bicarbonate and phosphate buffers. The bicarbonate ion, however, has proved to be an accurate and easily determined index of acid-base balance. The mechanism of buffer action may be understood from consideration of the bicarbonate system in eqn (4.1):

$$CO_2 + H_2O \leftrightharpoons H_2CO_3 \leftrightharpoons H^+ + HCO_3^- \quad (4.1)$$

The pH of blood can be influenced greatly by the level of CO_2. This may be seen by referring to eqn (4.1). If CO_2 is added or removed, the concentration of H^+ and HCO_3^- changes in the same direction. The pH-bicarbonate system is actually a representation of the Henderson-Hasselbach equation:

$$pH = pK + \log \frac{[HCO_3^-]}{[H_2CO_3]} \tag{4.2}$$

Clinically, four variations from the normal buffer point represent four types of acid-base disturbance: respiratory acidosis (excess CO_2), respiratory alkalosis (CO_2 deficit), metabolic acidosis (fixed acid excess), and metabolic alkalosis (fixed base excess). The kidneys are an important mechanism for base conservation. This is accomplished by acidification of urine and ammonia synthesis.

One of the main features of shock, for example, is respiratory and metabolic acidosis, in which CO_2 elimination by the lungs is inadequate; the PCO_2 of the alveoli and arterial blood is therefore increased, with a concomitant decrease in the pH (see eqns 4.1 and 4.2). Changes brought about in the buffer system by the infusion of hydroxyethyl starch will now be examined.

4.3.5.1. ANIMALS

The pH of dog arterial blood remains at physiological levels if serially diluted (1:2, 1:4, 1:8, and 1:16) *in vitro* with HES 450/0.70 (Takaori 1966).

In either dogs (Lindblad 1970) or rabbits (Irikura *et al.* 1972*g*, *h*, *u*) not subjected to haemorrhagic shock or haemodilution, the pH of arterial blood was not altered after these animals received large quantities of either HES 40/0.55 or HES 140/0.65 daily for up to three months.

In dogs administered either 3 or 6 g HES 40/0.55 kg in 1 hour, extravascular lung water increased concomitantly with an increase in arterial PCO_2 and a decrease in blood pH (see eqns 4.1 and 4.2) (Imazu *et al.* 1976, 1977).

In various dog and cat models producing acute haemorrhagic or burn shock or severe haemodilution, large increases in arterial PCO_2 concomitantly with changes in arterial or venous pH, standard bicarbonate, base excess, and buffer base are observed after the experimental treatment. A return to normal physiological levels for all parameters mentioned above occurs after resuscitation of animals with either HES 40/0.55 (Irikura *et al.* 1972*f*; Nakanishi 1972*a*, *b*; Saito 1972) or HES 450/0.70 (Ikeda *et al.* 1971; Jesch *et al.* 1975; Lee *et al.* 1965; Takaori *et al.* 1968; Thompson and Walton 1964*b*; Vineyard *et al.* 1966; Yoshikawa *et al.* 1974; Zimmermann and Bannert 1970). The return to normal values in these treated animals depends on several factors: (i) the intensity and duration of the performed experimental procedure (the more severe the procedure, the longer it takes for values to return to normal); (ii) the time between performing the experimental procedure and beginning resuscitation (abnormal blood levels persist longer in animals not resuscitated immediately after the experimental procedure); (iii) the intravascular survival of the injected species of hydroxyethyl starch (more slowly metabolized species of hydroxyethyl starch maintain flow and sustain plasma volume increases better than rapidly metabolized species (see Section 4.6.1 and Table 4.9).

4.3.5.2. MAN

In patients that had either sustained some degree of blood loss or were undergoing normovolaemic haemodilution, normal levels of arterial PCO_2 as well as arterial pH, buffer base, standard bicarbonate, base excess, and actual bicarbonate were maintained after the infusion of HES 40/0.55 (Hayashi and Higashi 1975), HES 200/0.60 (Nakajo 1972), or HES 450/0.70 (Hayashi and Higashi 1975; Kobayashi *et al.* 1971; Lee *et al.* 1968; Okada *et al.* 1971; Solanke 1968*a*; Watzek *et al.* 1978).

4.3.6. BLOOD GASES

When oxygen diffuses into the plasma from the lung alveolus, almost all of it finds its way into the erythrocyte, where it combines with haemoglobin. When the concentration of oxygen in simple solution or the PO_2 of the blood is low, the oxygen content of haemoglobin will also be low. In conditions where the PO_2 is increased, the oxygen content of haemoglobin will be high. The portion of the oxygen–haemoglobin dissociation curve of greatest importance is the *shoulder* in the PO_2 range of 40–100 mm Hg. At the top of this shoulder, large changes in PO_2 have a small effect on the amount of oxygen carried by haemoglobin; below this level, small changes in PO_2 have a large effect on the oxygen content. At the upper end of the dissociation curve, a decrease in alveolar oxygen tension does not immediately jeopardize the adequacy of the arterial oxygen content. On the other hand, at the lower end of the curve a decrease in venous oxygen content makes considerably more oxygen available to the tissues. The amount of oxygen combined with haemoglobin also depends upon other factors, such as the PCO_2 of blood. As the carbon dioxide tension increases, oxygen and haemoglobin tend to dissociate. Changes in PCO_2 after resuscitation with hydroxyethyl starch is discussed in Section 4.3.5. The oxygen–haemoglobin dissociation curve also depends upon the pH of blood; an increase in acidity tends to drive oxygen off haemoglobin (see Section 4.3.5 for discussions of the pH). In addition, the temperature of blood has a significant effect on oxygen carriage, an increased temperature causing dissociation of oxygen from haemoglobin.

The changes in blood gases (with the exception of PCO_2 – see Section 4.3.5) after resuscitation of animals and man with hydroxyethyl starch will now be discussed.

4.3.6.1. PO_2

Animals

Arterial PO_2 (P_aO_2) decreased 5 and 20 per cent after unbled dogs received 3 or 6 g/kg HES 40/0.55, respectively (Imazu *et al.* 1976). The reduction of P_aO_2 coincided with a significant increase in extravascular lung water. Similar observations were seen in unbled dogs receiving HES 40/0.55 in which intermittent positive pressure ventilation was used whilst maintaining the pulmonary capillary wedge pressure at 15 mm Hg (Imazu *et al.* 1977).

In dogs bled 30 ml/kg whole blood and then resuscitated 1–2 hours later with 1.8–2.7 g/kg HES 40/0.55, P_aO_2 levels rose after infusion (Katsuya *et al.* 1973; Satio 1972). In dogs undergoing isovolaemic haemodilution with HES 40/0.55 there was a direct correlation between a reduction in the haematocrit and decreased central venous PO_2 ($P_{\bar{v}}O_2$) levels (Messmer and Jesch 1978).

In dogs exchange-transfused down to a haematocrit of 10 per cent with HES 450/0.70, the level of P_aO_2 remained unchanged or slightly increased while $P_{\bar{v}}O_2$ was reduced (Jesch *et al.* 1975; Yoshikawa *et al.* 1974). In cats or dogs undergoing shock induced by the Wigger's method, PO_2 levels were slightly lower in cats (Zimmermann and Bannert 1970) but increased in dogs (Vineyard *et al.* 1966) 1 hour after resuscitation with HES 450/0.70. The elevated levels of PO_2 in dogs persisted for up to 20 hours after resuscitation.

Man

In hypovolaemic patients administered 30 g HES 450/0.70, P_aO_2 and $P_{\bar{v}}O_2$ after injection were significantly increased at their maximum level over pre-infusion values (Lazrove *et al.* 1980). The infusion of albumin under identical conditions produced similar increments in P_aO_2 and $P_{\bar{v}}O_2$.

In patients exchange-transfused with HES 450/0.70 down to a haematocrit of 26–28 per cent, P_aO_2 was increased after the end of the operation, while $P_{\bar{v}}O_2$ remained unchanged (Watzek *et al.* 1978). The venous and arterial P_{50} also remained essentially unchanged. In additional patients receiving HES 450/0.70 during resuscitation, PO_2 remained unchanged or slightly increased after the operation (Okada *et al.* 1971).

4.3.6.2. OXYGEN SATURATION

Animals

In dogs bled 30 ml/kg whole blood and resuscitated with 1.8 g/kg HES 40/0.55, arterial oxygen saturation (S_aO_2) initially increased immediately after infusion and then returned to prebled levels after 5 hours (Irikura *et al.* 1972*f*). In dogs exchange-transfused with HES 450/0.70 down to a haematocrit of 10 per cent, S_aO_2 remained low (Jesch *et al.* 1975) or unchanged (Safar *et al.* 1978; Takaori *et al.* 1968; Takaori and Safar 1976), while mixed venous oxygen saturation (S_vO_2) remained low 2 hours after termination of haemodilution. Twenty-four hours after initiation of the experimental procedure, S_vO_2 remained low.

Man

In hypovolaemic patients given 30–240 g HES 450/0.70, oxygen saturation after transfusion was unaltered in comparison with levels before infusion (Lazrove *et al.* 1980; Lee *et al.* 1968).

4.3.6.3. ALVEOLAR-ARTERIAL PO_2 DIFFERENCE

In dogs receiving 3 or 6 g/kg HES 40/0.55, alveolar-arterial PO_2 difference ($A{-}_aDO_2$) increased after infusion (Imazu *et al.* 1976). $A{-}_aDO_2$ also increased in dogs undergoing haemodilution with HES 450/0.70 (Yoshikawa *et al.* 1974). However, in dogs bled 30 ml/kg whole blood and then infused 2 hours later with 2.7 g/kg HES 40/0.55, the $A\text{-}_{a}DO_2$ *decreased* (Katsuya *et al.* 1973).

4.3.6.4. ARTERIO-VENOUS O_2 CONTENT DIFFERENCE

In dogs exchange-transfused with HES 450/0.70 down to a haematocrit of 10 per cent, the arterio-venous O_2 content difference ($C_{(a-v)}O_2$) decreased significantly during and immediately after haemodilution (Takaori *et al.* 1968; Takaori and Safar 1976). In dogs administered either dextran 40 or dextran 75 under identical test conditions, the decrease in $C_{(a-v)}O_2$ was more pronounced. Twenty-four hours after haemodilution was begun, $C_{(a-v)}O_2$ had returned to preinfusion values in all groups of dogs.

4.3.6.5. CAPILLARY O_2 CONTENT

In dogs bled 30 ml/kg whole blood and resuscitated with 2.7 g/kg HES 40/0.55 2 hours later, the arterial and mixed venous capillary oxygen content was reduced if determined 1 hour after completion of the infusion (Katsuya *et al.* 1973).

4.3.6.6. BODY TEMPERATURE

As discussed earlier in this section, the temperature of blood has a significant effect on oxygen carriage. Even although the temperature of blood after resuscitation with hydroxyethyl starch has not been directly determined, changes in body temperature after infusion may be useful to know.

No change in body temperature was observed in dogs given 3 g/kg HES 40/0.55 in 1 hour (Imazu *et al.* 1976). In dogs given large amounts of HES 40/0.55 (Imazu *et al.* 1976, 1977) the body temperature was reduced by an increment of up to 1.7°C.

In patients administered 30 g HES 450/0.70, the body temperature rose slightly (0.25 °C) 3 hours after the period of infusion (Lazrove *et al.* 1980).

4.3.7. OSMOTIC PRESSURE AND OSMOLARITY

In simple terms, osmotic pressure can be defined as the pressure required to prevent osmotic flow of water into a given solution. Osmotic pressure is one of the so-called colligative properties of dilute solutions, i.e. its magnitude is related to the concentration (number/unit volume) of dissolved particles and is not affected (at least in very dilute solutions) by such factors as particle size, shape, or chemical composition. Other colligative properties of dilute solutions are vapour pressure lowering, depression of the freezing point, and elevation of the boiling point. In most circumstances, osmotic pressure is directly proportional to solute concentration and for dilute solutions can be defined by the Van't Hoff equation:

$$\pi = CRT, \qquad (4.3)$$

where C is the solute concentration, R is the gas constant, and T is the absolute temperature. The solute has no intrinsic effect on osmotic pressure, that is, the osmotic pressure of a solution is independent of the nature of the solute particles and depends only on their number per unit volume. It is apparent with ionising solutes, such as electrolytes, however, that to relate the osmotic pressure or effective osmotic concentration of solute to its chemical concentration, it is necessary to multiply the term C in eqn (4.3) by a factor G (the osmotic coefficient), where G is the number of ions produced by one molecule of electrolyte. With solutions, such as blood or urine, which contain complex mixtures of solutes (both electrolytes and non-electrolytes) in widely different concentrations, it is necessary for a practical unit of osmotic concentration (which is independent of the way in which G varies with concentration for different solutes) to be defined. The *osmole* is such a unit. For osmotic purposes, the concentration of any given solution can be expressed directly in terms of osmolarity (number of *osmoles*/litre of solution). The following relationships, therefore, can be expressed for osmolarity, freezing point depression, and osmotic pressure:

$$\text{milliosmole} = \frac{T_f \times 1000}{1.86}, \tag{4.4}$$

$$\text{osmotic pressure (in atmospheres)} = \frac{T_f \times 22.4}{1.86}, \tag{4.5}$$

where T_f is the difference between the freezing point of pure water (0 °C) and that of the solution, 1.86 (1.86 °C/mole) the molecular depression of the freezing point and 22.4 is the ideal gas volume at one atmosphere. The change in both osmotic pressure and osmolarity in blood (see Section 4.4.3 for changes in voided urine) after the administration of various species of hydroxyethyl starch will now be described.

4.3.7.1. COLLOID OSMOTIC PRESSURE

The colloid osmotic pressure (COP) of human blood and plasma are 38.4 and 43.0 cm H_2O, respectively (Yamasaki 1973). Under static conditions *in vitro* the relationships between COP, colloid concentration, and water-binding ability for various species of hydroxyethyl starch (see Table 2.1 for physicochemical characteristics) and dextran (see Table 1.10 for physicochemical characteristics) are listed in Table 4.7. As is clearly shown, smaller quantities of osmotically active small molecules are required to achieve a similar COP as plasma. At similar molar concentrations, the COP is greater with small molecules than with larger species (Table 4.8). Therefore, as stated earlier, the osmotic pressure is directly related to the number of molecules per unit volume (eqn 4.3) and inversely related to molecular size (see Section 1.4.3).

Under conditions *in vivo*, the rate of formation of small molecules of hydroxyethyl starch depends on the MS of the injected species (Section 2.2.2.2).

TABLE 4.7 *Relationship of colloid osmotic pressure (COP) to colloid concentration.* *

Colloid	Colloid isotonic concentration (%)	COP (cm H_2O)	H_2O-binding ability (ml/g)
Dextran 40	2.7	41.2	37
Dextran 70	3.5	43.4	29
HES 40/0.55	3.7	41.5	27
HES 450/0.70	5.2	44.7	20
Whole blood	–	38.4	–
Plasma	–	43.0	–
Albumin	–	–	19

* From Yamasaki (1973)

Species of hydroxyethyl starch possessing an MS of less than 0.60 are catabolized to smaller molecules sooner than species with an MS over 0.60. In terms of COP and water-binding ability, species of hydroxyethyl starch that are metabolized rapidly produce a greater number of molecules per unit volume and are, therefore, more osmotically active. This point will be explained further in relation to changes in plasma volume after infusion (see Section 4.6.1).

TABLE 4.8 *Relationship of COP to a 6% (w/v) Solution of Various Colloids (Yamasaki, 1973).*

Colloid	COP (cm H_2O)
Dextran 40	125.0
Dextran 70	108.1
HES 40/0.55	77.6
HES 450/0.70	58.5

In rabbits or dogs, the COP is increased after infusion of either HES 40/0.55 (Fukutome 1977; Yamasaki 1975) or HES 450/0.70 (Fukutome 1977; Jesch *et al.* 1975). In dogs it was shown that a similar increase in the COP was achieved by an intravascular plasma concentration of 5 mg/ml HES 40/0.55 and 10 mg/ml HES 450/0.70 (Fukutome 1977). These data confirm the *in vitro* data presented in Table 4.7. These data also demonstrate that the osmotic activity of HES 40/0.55 is approximately twice that of HES 450/0.70 (Table 4.8). Even although HES 40/0.55 is more osmotically active than HES 450/0.70, the intravascular survival of this smaller species of hydroxyethyl starch is approximately one-tenth that of HES 450/0.70 (Table 4.9).

In studies on the dog performed by Messmer and Jesch (1978), there appeared to be a direct relationship between COP and the degree of isovolaemic haemodilution produced by HES 40/0.55. As the haematocrit fell, there was a concomitant decrease in COP.

TABLE 4.9 *Maintenance of an increased plasma volume with the persistence of hydroxyethyl starch*

Species	Intravascular half-life (hours)	Maximum H_2O-binding *in vivo* (ml/g)	Colloid osmotic pressure (cmH_2O)	Return to normovolaemia after injection (hours)	Reference
HES 40/0.55	2.9	27	77.6	1.5	1,5,6,8,9
HES 264/0.43	2.7	27–33	66.4	24	1,3,4
HES 450/0.70	33	19–21	58.5	72	1,2,7,8

1 – Yamasaki (1975); 2 – Metcalf *et al.* (1970); 3 – Mishler *et al.* (1979*b*); 4 – DiMarco *et al.* (1978); 5 – Kitamura *et al.* (1972*a*,*b*); 6 – Kori-Lindner and Hubert (1978); 7 – Hayashi and Higashi (1975); 8 – Yamasaki (1973); 9 – Nakanishi (1972*a*,*b*)

Information on the effect of various species of hydroxyethyl starch on COP in patients is rather limited. In two independent studies, it was shown that the COP was virtually unchanged after the infusion of either 30 g HES 40/0.55 (Grünert *et al.* 1978) or 60 g HES 450/0.70 (Jesch *et al.* 1979; Watzek *et al.* 1978).

4.3.7.2. OSMOLALITY

The osmolalities of various hydroxyethyl starch species are given in Table 2.1. All species of hydroxyethyl starch are suspended in 0.9 per cent isotonic saline with the exception of HES 40/0.55, which is suspended in Ringer's lactate.

In dehydrated rabbits administered 3.6 g HES 40/0.55/kg (Yamasaki 1975) or in dogs given 1.2–1.8 g HES 40/0.55/kg (Fukutome 1977; Sudo *et al.* 1972), the osmolality of plasma fell after infusion. The reduction in osmolality was not pronounced (– 5 mOsmol/l) in dogs receiving 1.2 g/kg (Fukutome 1977), but marked decreases (– 26 to – 30 mOsmol/l) were observed in rabbits or dogs administered more than 1.8 g/kg. Plasma osmolality was essentially unchanged in dogs given 1.2 g/kg HES 450/0.70 (Fukutome 1977). However, in dogs exchange-transfused with HES 450/0.70 down to a haematocrit of 10 per cent, plasma osmolality remained decreased (21 mOsmol/l below control value) nine days after exchange (Jesch *et al.* 1975).

In patients or normal subjects administered 30–60 g of HES 40/0.55 (Kitamura *et al.* 1972*a*, *b*), HES 200/0.60 (Nakajo 1972; Yamamoto *et al.* 1974), or HES 450/0.70 (Kilian *et al.* 1975; Kobayashi *et al.* 1971; Watzek *et al.* 1978), plasma osmolality was essentially unchanged after injection.

In normal subjects administered 30 g HES 450/0.70 on four occasions during a 12-day interval (total 120 g), plasma osmolality remained fairly constant (Maguire *et al.* 1981; Strauss and Koepke 1979). These data complement the findings presented in Section 4.3.3 that plasma electrolytes are not greatly influenced by the infusion of any species of hydroxyethyl starch.

4.3.8. WHOLE BLOOD AND PLASMA VISCOSITY

Human plasma is approximately 1.8 times as viscous as water. Whole blood has a variable viscosity ranging between two and fifteen times that of water. One of the most important factors influencing viscosity and flow of whole blood is the haematocrit. In most circumstances, the viscosity is directly proportional to the haematocrit.

At low and near-zero flow conditions, the viscosity of blood is high and strongly related to the shear rate which is low (Gregersen *et al.* 1965). Measurements of viscosity at known low shear rates are potentially important because in shock and other forms of circulatory failure with stagnation and sluggish flow in the microcirculation, presumably the flow rates are near zero. Gregersen and his colleagues (1965) have shown, for example, that the viscosity of whole blood is inversely proportional to shear rate. A similar relation is observed when hydroxyethyl starch is added to whole blood. These same investigators substantiated a general correlation between increase in plasma viscosity and colloid (hydroxyethyl starch and dextran) molecular weight, even although hydroxyethyl starch has a comparatively low viscosity in relation to its molecular weight (see Section 1.4.3). Additional studies (Messmer and Jesch 1978) have shown a direct correlation between relative viscosity (apparent blood viscosity/plasma viscosity) and degree of haemodilution with HES 450/0.70. In this example, the relative viscosity rose as the haematocrit increased. In this same study, Messmer and Jesch demonstrated that the viscosity of plasma fell as the haematocrit rose. The effect of hydroxyethyl starch on whole blood and plasma viscosity *in vivo* will now be examined. In addition to the effect of hydroxyethyl starch on viscosity, it must be remembered that dilutional effects are also present (see Sections 4.3.1 and 4.3.2) in as much as the concentration of erythrocytes and plasma proteins will be reduced in direct proportion to the quantity of hydroxyethyl starch infused.

4.3.8.1. ANIMALS

In dogs exchange-transfused with HES 40/0.55 down to a haematocrit of 10 per cent (Nakanishi 1972*a*) or dogs resuscitated with HES 40/0.55 after being bled 25 ml/kg whole blood (Mori 1975; Nakanishi 1972*b*), the viscosity index (viscosity/haematocrit) was not markedly changed. In dehydrated rabbits given 3.6 g/kg HES 40/0.55, whole-blood viscosity at low shear rates was reduced markedly 1 hour after transfusion (Yamasaki 1975). This reduction in blood viscosity persisted 12 hours or more. In dogs exchange-transfused with HES 450/0.70 down to a haematocrit of 10 per cent, the apparent viscosity of whole blood at both low and high rates of shear, as well as the relative viscosity, were all reduced 40–46 per cent from prebled levels (Jesch *et al.* 1975).

4.3.8.2. MAN

The relative viscosity of both whole blood and plasma was reduced by 10 and 4 per cent, respectively, in normal subjects given a single 30 g infusion of HES 40/0.55 (Ehrly *et al.* 1979). The reduction in viscosity was maintained for approximately 3 hours after infusion. In subjects given two 30 g injections of

HES 40/0.55 within 2 hours, the relative viscosities of both whole blood and plasma were reduced, even though the effect of these two injections was not additive (Ehrly *et al.* 1979). In patients given 60 g HES 450/0.70 and 500 ml 3.6 per cent albumin, the viscosity of both whole blood and plasma, in addition to the relative viscosity of both low and high rates of shear had been reduced by 20–40 per cent relative to preinfusion values 16 hours after injection (Watzek *et al.* 1978). In patients receiving either 30-90 g (9-12 g/h) or 120-240 g (0.45 g/kg per h) of HES 450/0.70, the apparent blood viscosity, in general, was somewhat lower in the group of patients who received higher doses (Lee *et al.* 1967, 1968).

4.4. CHANGES IN URINE CONSTITUENTS AFTER RESUSCITATION

The effect of various species of hydroxyethyl starch on renal function has been presented in Section 3.9.2.

4.4.1. ELECTROLYTES

4.4.1.1. SODIUM

In dogs bled 30 ml whole blood/kg and resuscitated with 1.8 g/kg HES 40/0.55 (Irikura *et al.* 1972*f*), in rabbits given over 1.8 g/kg HES 450/0.55 daily for one month (Irikura *et al.* 1972*l*) and in dehydrated rabbits administered 3.6 g/kg HES 40/0.55 (Yamasaki 1975), the urinary excretion of sodium was increased over preinfusion levels. In those studies in which HES 40/0.55 was acutely administered, the enhanced excretion was temporary (1-3 hours), even although in dehydrated rabbits a second period of increased excretion was observed 24 hours after injection (Yamasaki 1975). In dogs bled 50 per cent of their blood volume and resuscitated with HES 450/0.70, a reduction in the excretion of sodium was observed (Heidenreich *et al.* 1975).

The excretion of sodium was increased over control levels in patients administered either 30 g HES 40/0.55 (Kitamura *et al.* 1972*a*,*b*) or 60 g HES 450/0.70 (Hempel *et al.* 1975; Kobayashi *et al.* 1971). Twenty-four hours after injection, these enchanced values returned to normal.

4.4.1.2. POTASSIUM

In bled dogs resuscitated with either HES 40/0.55 (Irikura *et al.* 1972*f*) or HES 450/0.70 (Heidenreich *et al.* 1975), the excretion of potassium was slightly increased after infusion of HES 40/0.55 and in the case of HES 450/0.70, the elimination was reduced slightly. In dehydrated rabbits given HES 40/0.55, a pronounced excretion of potassium was observed 12-24 hours after injection (Yamasaki 1975). Increased quantities of potassium appeared in the urine of rabbits administered over 1.8 g/kg HES 450/0.55 daily for one month (Irikura *et al.* 1972*l*).

A slight increase in the excretion of potassium was observed in patients given either 30 g HES 40/0.55 (Kitamura *et al.* 1972*a,b*) or 60 g HES 450/0.70 (Hempel *et al.* 1975; Kobayashi *et al.* 1971).

4.4.1.3. CHLORIDE

The excreted levels of chloride were not altered in dogs bled 30 ml/kg whole blood and resuscitated with 1.8 g/kg HES 40/0.55 (Irikura *et al.* 1972*b*). However, the excretion of chloride in dehydrated rabbits given 3.6 g/kg HES 40/0.55, rose 1 hour after infusion and again at 24 hours (Yamasaki 1975).

The level of excreted chloride was increased in patients administered 60 g HES 450/0.70 (Hempel *et al.* 1975; Kobayashi *et al.* 1971).

4.4.1.4. CALCIUM

Excretion of calcium in patients resuscitated with 60 g HES 450/0.70 was temporarily reduced after the injection but returned to preinfusion levels by 24 hours (Kobayashi *et al.* 1971).

4.4.2. pH

In rabbits administered more than 1.8 g/kg HES 40/0.55 daily for up to three months, the pH of excreted urine was not significantly changed in relation to intact controls (Irikura 1972; Irikura *et al.* 1972*g*). This observation was also true in rabbits receiving over 1.8 g/kg HES 450/0.55 daily for one month (Irikura *et al.* 1972*l*).

4.4.3. COLLOID OSMOTIC PRESSURE AND OSMOLARITY

4.4.3.1. COLLOID OSMOTIC PRESSURE

In dogs given 1.2 g/kg HES 40/0.55 (Fukutome 1977) and in dehydrated rabbits administered 3.6 g/kg HES 40/0.55 (Yamasaki 1975), the colloid osmotic pressure (COP) of urine rose significantly after infusion, returning to preinfusion levels after 24 hours. In dogs receiving 1.2 g/kg HES 450/0.70, the COP rose after injection to a greater degree than an equal injection of HES 40/0.55 (Fukutome 1977). Six hours after injection, however, the level of COP was similar for both colloids.

4.4.3.2. OSMOLARITY

In dehydrated rabbits (Yamasaki 1975) or normal dogs (Fukutome 1977; Sudo *et al.* 1972) receiving over 1.2 g/kg HES 40/0.55, the osmolarity of urine decreased after injection, but returned to approximate preinfusion levels by 24 hours. In dogs injected with 1.2 g/kg HES 450/0.70, the osmolarity of urine rose slightly or was decreased after infusion (Fukutome 1977; Murphy *et al.* 1965). Post-injection urine osmolarity fell, however, in dogs resuscitated with HES 450/0.70 after haemorrhagic shock was induced by a two-stage Wigger's technique (Vineyard *et al.* 1966).

The osmolarity of urine in patients receiving 30 g HES 40/0.55 increased after injection but returned to approximate preinfusion levels by 24 hours (Grünert *et al.* 1978; Kitamura *et al.* 1972*a,b*). In normal subjects injected with 60 g HES 200/0.65, the osmolarity of urine decreased initially 1 hour after injection but rose above control levels 3 hours after infusion (Yamamoto *et al.* 1974). In patients receiving the same quantity of HES 200/0.65 as mentioned above, the osmolarity of urine initially rose after injection, but fell below preinjection levels 7 hours after dosing (Nakajo 1972). In patients dosed with 60 g HES 450/0.70, the osmolarity of urine rose after injection and remained increased 24 hours after infusion (Kobayashi *et al.* 1971).

4.4.4. SPECIFIC GRAVITY AND VISCOSITY

4.4.4.1. SPECIFIC GRAVITY

In Section 2.2.3 (Table 2.2) it was shown that the amount of hydroxyethyl starch appearing in urine after infusion was an inverse property of the $M_{\bar{n}}$. Thus, as a function of time after dosing, a greater proportion of the total amount of HES 40/0.55 injected will be excreted compared with HES 450/0.70. Even although this observation has been documented, the infusion of equal doses of HES 40/0.55 and HES 450/0.70 under identical test conditions in dogs (Fukutome 1977) or man (Hayashi and Higashi 1975) will produce similar increments in the specific gravity of voided urine. Additional studies in dehydrated rabbits have shown that the specific gravity of urine rises initially after an injection of 3.6 g/kg HES 40/0.55 but returns to preinfusion levels after 24 hours (Yamasaki 1975). In circumstances where rabbits are administered more than 1.8 g/kg HES 40/0.55 daily for 1 month, the specific gravity of urine is virtually unaffected (Irikura *et al.* 1972*g*).

4.4.4.2. VISCOSITY

In dogs given equal doses of either HES 264/0.43 or dextran 40 under identical test conditions, it was shown that dextran 40 significantly increased the viscosity of voided urine, whereas HES 264/0.43 did not, even although larger quantities of HES 264/0.43 appeared in urine as time elapsed (Thompson *et al.* 1977*a,b*). In dehydrated rabbits administered equal doses of either HES 40/0.55 or dextran 40 under identical test conditions, the viscosity of urine was significantly increased in the dextran 40 group (Yamasaki 1975). In this situation, however, an increasing proportion of the infused dextran 40 appeared in voided urine as time elapsed.

In normal subjects dosed with either HES 200/0.50 (Mishler 1980*b*) or HES 264/0.43 (Mishler *et al.* 1980*d*), the relative viscosity of urine measured 1 hour after administration of either colloid was unaffected, even although 15 per cent of the total injected dose appeared in urine during this period.

4.5. VISCERAL BLOOD FLOW AFTER RESUSCITATION

4.5.1. RENAL BLOOD FLOW

The direct infusion of HES 450/0.70 (0.18 g/min) into the renal artery of dogs resulted in an increase in direct renal blood flow (DRBF) and renal venous pressure with a concomitant decrease in renal resistence (Silk 1966). Results obtained after diphenhydramine hydrochloride pretreatment, and/or after HES 450/0.70 was incubated with blood before infusion did not vary significantly from the results when HES 450/0.70 was injected alone.

In an acute normotensive dog preparation, no augmentative benefit in DRBF was observed after the infusion of 1.2 g/kg HES 450/0.70 (Murphy *et al.* 1965). However, an increase in renal blood flow (RBF), a decrease in plasma total solids and in peripheral arterial haematocrit value, with maintenance of normal blood pressure, were observed in these circumstances. The authors concluded that the failure to achieve a rise in DRBF under these conditions could be attributed to the animal preparation used. In intact or splenectomized dogs receiving 3.9 or 6.1 g/kg HES 450/0.70, a general increase in RBF was associated with a decreasing arterial haematocrit (Murphy 1965*a, b*). In intact or splenectomized dogs receiving 5.5 or 5.9 g/kg HES 450/0.70, after an injection of phenoxybenzamine, RBF was diminished in response to haemodilution. In dogs subjected to acute haemorrhagic shock in which the total quantity of blood withdrawn was replaced with HES 450/0.70, the RBF was increased by 50 per cent over preshock levels (Heidenreich *et al.* 1975).

No significant changes were noted in RBF in rabbits administered 0.6 or 1.8 g/kg per day HES 40/0.55 for ten days (Irikura *et al.* 1972*b*). A significant increase in RBF was observed in rabbits given 5.4 g/kg per day HES 40/0.55 for ten days.

4.5.2. MESENTERIC BLOOD FLOW

In dogs bled 30 ml/kg whole blood and resuscitated with 1.8 g/kg HES 40/0.55, the mesenteric blood flow was increased over preinfusion levels (Irikura and Kudo 1972*a*).

4.5.3. FLOW IN THE PORTAL VEIN

In cats undergoing shock induced by the Wigger's technique, replacement of the blood lost by an equi-volume infusion of HES 450/0.70 resulted in a significant increase of flow in the portal vein, 1 hour after resuscitation (Zimmermann and Bannert 1970).

4.5.4. MISCELLANEOUS

In dogs undergoing haemodilution with HES 450/0.70, blood flow to all organs increased during haemodilution due to an increase in cardiac output (Yoshikawa *et al.* 1974). Blood supply increased in the heart, skin, pancreas, and brain, remained unchanged in muscle, liver, and gastrointestinal tract, and decreased in kidney and adrenal glands.

4.6. CHANGES IN BODY FLUID COMPARTMENTS

The fate of extravasated hydroxyethyl starch (Section 2.3) and its appearance in lymph (Section 3.10) have been discussed. Thus, the movement of hydroxyethyl starch from the intravascular space into both extracellular and interstitial spaces would be expected to alter body fluids because of the colloid osmotic pressure gradient exerted by the movement of this material (see Section 4.3.7). Changes in plasma volume (PV) and concomitant alterations in the extracellular (ECF), intracellular (ICF), and interstitial (ISCF) fluid volumes will now be discussed.

4.6.1. PLASMA VOLUME

The overall increment in the PV after the administration of hydroxyethyl starch will depend on several factors: the amount of colloid given; the concentration (weight/volume) of the injected solution, and the molecular weight of the injected species of hydroxyethyl starch (colloid osmotic pressure is inversely related to molecular weight; Section 1.4.3). The maintenance of the enhanced PV after infusion of hydroxyethyl starch depends on the persistence of this material in blood (persistence is directly related to hydroxyethyl group substitution (MS); Section 2.2.2).

Several general conclusions on the increment in the PV after infusion of hydroxyethyl starch can be stated, based on studies in animals (Fukutome 1977; Gollub *et al.* 1967, 1969*a*; Ikeda *et al.* 1971; Irikura *et al.* 1972*f*, *g*; Jesch *et al.* 1975; Messmer and Jesch 1978; Safar *et al.* 1978; Sudo *et al.* 1972; Takaori and Safar 1966, 1967, 1976; Takaori *et al.* 1970; Thompson and Walton 1964*b*) and man (Ballinger *et al.* 1966*a*; DiMarco *et al.* 1978; Fukuda 1972; Goto *et al.* 1972*b*; Hempel *et al.* 1975; Kilian *et al.* 1975; Kinoshita 1971; Köhler *et al.* 1978*a*, *b*; Kori-Lindner and Hubert 1978; Lamke and Liljedahl 1977; Lazrove *et al.* 1980; Lee *et al.* 1968; Metcalf *et al.* 1970; Mishler *et al.* 1979*b*; Solanke *et al.* 1971; Ulmi 1979). Firstly, the infusion of a 6 per cent solution of either HES 40/0.55 or HES 450/0.70 will increase the PV slightly *less* or *equal* to the volume of colloid administered. This observation would seem surprising for HES 40/0.55 if it were not for the fact that this species of hydroxyethyl starch is catabolised quite rapidly (40 per cent of the injected dose remains in blood after 5 hours) (Kitamura *et al.* 1972*a*,*b*), with large quantities appearing in voided urine (Fig. 2.8). HES 450/0.70, on the other hand, has the same water-binding capacity as albumin and persists longer in blood. Secondly, the infusion of either a 10 or 14 per cent solution of HES 264/0.43 (HES 200/0.50) will increase the PV 1.5 to 1.8 times the amount infused. This species of hydroxyethyl starch has a lower molecular weight than HES 450/0.70, and low molecular weight material is continually produced after infusion (Section 2.2.2).

The maintenance of the enhanced PV after infusion is directly related to the post-transfusion survival in blood of the injected species of hydroxyethyl starch (Table 4.9). Even although the intravascular half-lives of HES 40/0.55 and HES 264/0.43 are similar, HES 264/0.43 persists in blood longer, accounting for the longer time required to return to normovolaemia. HES 450/0.70, on the other hand, is catabolized slowly, resulting in a more persistent increase of the PV after infusion. The relationship between persistence in blood and the concomitant increment in the PV can be illustrated by an additional example. In patients with terminal renal failure, the increment in PV after the infusion of HES 450/0.70 is maintained longer when compared, under identical test conditions, with patients possessing normal renal function (Köhler *et al.* 1977*d*, 1978*b*). This observation is not surprising considering that the main mode of elimination of hydroxyethyl starch is excretion through the kidneys (Section 2.2.3), and that if excretion is hindered large quantities will persist in blood for greater periods of time.

4.6.2. EXTRACELLULAR FLUID VOLUME

The effect on ECF after the infusion of hydroxyethyl starch remains controversial.

In splenectomized dogs administered 1.8 g/kg HES 40/0.55, the PV and ECF *increased* 12 and 13 per cent, respectively, after infusion (Sudo *et al.* 1972).

In dogs exchange-transfused with HES 450/0.70 down to a haematocrit of 10 per cent, ECF *increased* 38 per cent immediately after the experimental procedure concomitantly with an increase of 100 per cent for the PV (Takaori *et al.* 1970). The ECF decreased towards control values after eight days. In dogs bled 30–35 per cent of their circulating blood volume and resuscitated immediately or 1 hour later with an equal volume of HES 450/0.70, ECF and PV *increased* 10 and 30 per cent, respectively, in both situations 1 hour after injection (Ikeda *et al.* 1971). HES 450/0.70, when diluted with Ringer's lactate solution and given in twice the quantity of shed blood, effectively restored the depleted ECF observed before fluid replacement (Gollub *et al.* 1969*a*).

In surgical patients given 6–45 g HES 450/0.70, ECF *decreased* by 14 per cent when compared with control values (Kinoshita *et al.* 1971). The ECF *increased*, however, after patients were administered HES 200/0.60 (Nakajo 1972).

4.6.3. INTRACELLULAR FLUID VOLUME

In splenectomized dogs injected with 1.8 g/kg HES 40/0.55, the ICF decreased by 14 per cent after dosing (Sudo *et al.* 1972).

In dogs exchange-transfused with HES 450/0.70 down to a haematocrit of 10 per cent, there was a significant deficit of ICF at the termination of haemodilution, but this returned to normal 2 hours later (Takaori and Safar 1967; Takaori *et al.* 1970).

4.6.4. INTERSTITIAL FLUID VOLUME

The ISCF *increased* 2 per cent after splenectomized dogs received 1.8 g/kg HES 40/0.55 (Sudo *et al.* 1972).

In dogs either exchange-transfused with HES 450/0.70 down to an haematocrit of 10 per cent (Takaori and Safar 1966, 1967; Takaori *et al.* 1970) or given an equal volume of HES 450/0.70 to replace 30–35 per cent of shed whole blood (Ikeda *et al.* 1971), ISCF was essentially *unchanged* after the experimental procedure was completed. On the other hand, dextran 40 *decreased* ISCF after infusion (Kinoshita *et al.* 1971; Takaori *et al.* 1970; Yamasaki 1974).

In patients dosed with 6–45 g HES 450/0.70, ISCF *decreased* by 17 per cent after infusion (Kinoshita *et al.* 1971). The ISCF increased slightly, however, in patients receiving HES 200/0.60 (Nakajo 1972).

4.6.5. LUNG WATER

In dogs given 3 or 6 g/kg HES 40/0.55, a significant increase in extravascular lung water was observed (Imazu *et al.* 1976).

4.6.6. SUMMARY

Most data presented in this section would appear to suggest that the initial expansion of the PV after the infusion of either HES 40/0.55 or HES 450/0.70 may be at the expense of ECF, since ISCF is essentially unchanged (with the exception of the data of Kinoshita and his colleagues (1971)) during this period. Perhaps intracellular dehydration occurs initially because small molecules of hydroxyethyl starch are not concentrated rapidly enough in cells after infusion to maintain a colloid osmotic pressure gradient across cell membranes. Additional studies may elucidate this observation further.

4.7. EFFICACY OF TREATMENT DURING SEVERE HAEMODILUTION OR HAEMORRHAGIC SHOCK

The effectiveness of resuscitation with hydroxyethyl starch can be measured by changes taking place in compensatory mechanisms after severe haemodilution or haemorrhagic shock.

Replacement of blood volume will generally improve perfusion of blood to vital organs when blood volume is deficient (Thompson 1974). In most animals or patients with hypoperfusion not related to blood loss or hypovolaemia, augmentation of blood volume and heart filling will increase cardiac output (CO) and vital organ perfusion. If blood volume augmentation is continued to high filling pressures, CO will no longer increase and oedema (both pulmonary and peripheral) will occur. Patients with hypoperfusion are managed by measuring heart filling pressure (central venous (CVP), pulmonary arterial diastolic (PAD), or pulmonary wedge (PWP) pressures) and vital organ perfusion (CO, mean arterial pressure (MAP), urinary output, and control pulse quality).

Two patterns of patient response to blood volume augmentation are normally seen (Thompson 1974). In patients with *high cardiovascular compliance*, an increment of 200 ml in blood volume will cause improved perfusion and a transient small increase in filling pressures that return to levels of under 15 torr. If perfusion remains inadequte, these patients should be administered additional blood volume increments. In patients with *low cardiovascular compliance*, blood volume increments do not improve perfusion and cause a persistent large increase in filling pressures to over 20 torr. In this latter group, continued blood volume augmentation may elicit pulmonary oedema, and therapy with inotropic drugs and vasodilators should be considered.

Right ventricular filling pressure (CVP) reflects left ventricular filling pressure in many patients. In some patients with left ventricular myocardial infarction, burns, and sepsis, left ventricular end-diastolic pressures (LVEDP) are much greater than right heart filling pressures, but on volume loading both pressures tend to change in the *same* direction and to the *same* extent. In some patients with cor pulmonale or right ventricular myocardial infarction, right ventricular filling pressures must be increased to levels of 30–40 torr to achieve adequate left ventricular filling. In these special circumstances, direct measurement of LVEDP or pulmonary arterial wedge pressure (PAWP) may be useful.

As ventricular filling pressures are increased, pulmonary oedema will ensue when pulmonary capillary hydrostatic pressures exceed effective plasma colloidal osmotic pressure. The choice of fluids for augmentation of blood volume is thus critical (Thompson 1974). The remaining portion of this section will specifically address the changes taking place in compensatory mechanisms after augmentation of the blood volume with hydroxyethyl starch.

4.7.1. SEVERE TO MODERATE HAEMODILUTION

In dogs exchange-transfused with HES 450/0.70 down to a haematocrit of 10 per cent, MAP decreased slightly after haemodilution and reached the lowest values 24 hours later (Jesch *et al.* 1975; Takaori and Safar 1965, 1967, 1976; Takaori *et al.* 1965, 1968, 1970, 1971; Yoshikawa *et al.* 1974). Total peripheral resistance (TAP) also decreased after haemodilution. The heart rate (HR), as well as right ventricular end-diastolic pressure (RVEDP), cardiac index (CI), and stroke index (SI) were essentially unchanged after haemodilution. The mean right atrial pressure (MRAP) as well as CVP rose after haemodilution, returning to normal levels 24 and 2 hours later. The QRS vector is the electrocardiogram (ECG) decreased during and after haemodilution. Premature ventricular contractions occurred occasionally.

In dogs undergoing isovolaemic haemodilution with HES 40/0.55, CO and coronary blood flow increased after haemodilution (Saito *et al.* 1976). The increase in CO was thought to result in a concomitant increase in left ventricular work. A decrease in TPR was also noted after haemodilution. The F cell ratio remained unchanged after haemodilution (Nakanishi 1972*a*, *b*).

In rats exchange-transfused with HES 450/0.70 down to a haematocrit of 20 per cent, MAP was essentially unchanged after haemodilution (Hölscher 1973/74; Hölscher *et al.* 1975).

4.7.1.1. MAN

In patients exchange-transfused with HES 450/0.70 down to a haematocrit ranging between 26 and 32 per cent, HR and MAP were essentially unchanged after haemodilution (Kraatz *et al.* 1975; Watzek *et al.* 1978). Increases in CO, CVP, stroke volume (SV), and pulmonary pressure were observed. A decrease in TPR was seen during haemodilution.

4.7.2. HAEMORRHAGIC SHOCK

The successful resuscitation and management of haemorrhagic shock depends on several factors. Firstly, the severity of the insult (amount of blood lost) must be considered. Normally an inverse relationship is observed between severity and successful treatment. Secondly, the time interval between insult and resuscitation plays a major role in management. An inverse relationship is often seen between insult and resuscitation. In this section I will attempt to document the severity of the insult and the time interval between the 'state of shock' and resuscitation.

4.7.2.1. ANIMALS

In normovolaemic dogs administered 1.2 g/kg HES 450/0.70 and *not* subjected to haemorrhagic shock, CVP and MAP rise after infusion and fall towards normal preinjection levels 2 hours later (Fukutome 1977).

In dogs maintained at 30–40 torr by withdrawal of blood and resuscitated immediately (Gollub *et al.* 1969*a*), 30 min (Thompson and Walton 1964*b*), 1 hour (Dillon *et al.* 1966; Vineyard *et al.* 1966) or 2 hours (Smith *et al.* 1975, 1977) later with HES 450/0.70, CO and HR increased moderately in all treatment groups after resuscitation. MAP and CVP were restored to near control values. Femoral arterial pulse pressures after infusion were greater than pre-haemorrhage values. TPR was decreased after resuscitation.

In cats maintained at 40–50 torr for 2 hours and then resuscitated with HES 450/0.70, CVP and CI rose above preshock levels, 1 hour after transfusion (Zimmermann and Bannert 1970). MAP and blood pressure (BP) were restored to near control levels, whilst portal venous pressure remained essentially unchanged.

In normovolaemic dogs administered 1.2 g/kg HES 40/0.55, the CVP rose immediately after infusion and returned to preinfusion levels 4 hours later (Fukutome 1977). Under these conditions, MAP was essentially unchanged after the injection.

In rabbits or dogs bled 10, 20, or 30 ml/kg whole blood and then resuscitated immediately with an equal volume of HES 40/0.55, HR, MAP, CVP, BP and

ECG patterns returned to preshock control levels after infusion (Irikura and Kudo 1972*b*; Irikura *et al.* 1972*e,f,g*; Saito 1972). MAP and CVP as well as BP, however, gradually declined thereafter when determined up to 5 hours after transfusion (Irikura *et al.* 1972*f*; Saito 1972). In studies using one species of hydroxyethyl starch in which the degree of MS was varied, BP was maintained in direct relationship to MS (Irikura and Kudo 1972*b*). In dogs maintained at 40 torr for 30 minutes and then resuscitated with HES 40/0.55, MAP, CVP, CO, and pulmonary pressure returned to preshock levels after infusion (Hartung *et al.* 1979). Pulmonary resistance decreased after infusion. These variables were maintained for 3–4 hours after infusion. In dogs bled 30 ml/kg whole blood and resuscitated 2 hours later with 1.8 g/kg HES 40/0.55, CO decreased to 30 per cent of control values after haemorrhage and returned to approximately 60-70 per cent of control values after infusion (Katsuya *et al.* 1973). MAP returned to preshock levels after the infusion of HES 40/0.55. A positive correlation was found between pulmonary shunt ratio and CO. Additional studies in animals have shown hydroxyethyl starch to be effective in restoring normal haemodynamic characteristics after shock (Saito *et al.* 1976; Suyama *et al.* 1971; Uemura 1973).

4.7.2.2. MAN

In normovolaemic subjects administered 30 g HES 450/0.70, CVP was increased after infusion while MAP and HR were essentially unchanged (Takeyoshi *et al.* 1971*a,b*).

The low BP induced by epidural anaesthesia returned to normal levels and was maintained after the infusion of 30 g HES 450/0.70 (Hayashi and Higashi 1975). In hypovolaemic patients given 30 g HES 450/0.70, MAP and CI were slightly increased after infusion (Lazrove *et al.* 1980). On the other hand, CVP, wedge pressure, and left and right ventricular stroke work indices were moderately increased after resuscitation. In the study by Lazrove and his colleagues (1980) and in an additional study (Munoz *et al.* 1980), HES 450/0.70 restored normal haemodynamic characteristics as well as albumin under similar test conditions.

In hypovolaemic patients given 60-90 g HES 450/0.70, CO, MAP, BP, CI, and SI rose above preinfusion levels after infusion (Hempel *et al.* 1975; Homann *et al.* 1977; Lee *et al.* 1968). HR remained essentially unchanged after infusion and a reduction in TRP was noted.

If given in approximately twice the volume of estimated blood lost, HES 200/0.60 serves to restore blood volume adequately with significant changes in BP and HR (Yamamoto *et al.* 1974).

In postoperative patients given a rapid infusion (30 g in 10 minutes) of HES 40/0.55, CI, SV, and CVP rose appreciably immediately after the infusion (Eisele *et al.* 1979). HR was essentially unchanged after infusion and a moderate decrease in TPR was noted.

In patients given 15 g HES 40/0.55 after open-heart surgery, CI as well as MAP, peripheral, pulmonary, and left atrial pressures increased after infusion, returning to control levels 2 hours later (Yamada and Sakamoto 1975). HR and SI were essentially unchanged after giving HES 40/0.55, and CVP rose initially but fell rapidly.

In hypovolaemic patients receiving 30 g HES 40/0.55, the induced low BP was restored to preshock levels after infusion (Kori-Lindner and Hubert 1978). Additional studies in patients administered 30 g HES 40/0.55 have shown that normal BP is *not* maintained well after infusion (Goto *et al.* 1972*b*; Hayashi and Higashi 1975). In the former study, HR was essentially unchanged after infusion, while in the latter studies, HR decreased after resuscitation. In studies where patients were administered 30 g HES 40/0.55 in addition to 500–1000 ml crystalloid solution, BP was slightly increased after infusion concomitantly with an unaltered HR (Kitamura *et al.* 1972*a, b*; Kori-Lindner and Hubert 1978). The balance between the quantity of blood lost and the amount of HES 40/0.55 infused has been studied in hypovolaemic patients (Kinoshita *et al.* 1971). The CVP was raised 6 per cent over preshock levels when the amount of blood lost was replaced with an equal quantity of HES 40/0.55. The CVP was raised 30 and 82 per cent when there existed a positive balance of 100 and 350 ml, respectively.

4.7.3. BURN SHOCK

4.7.3.1. ANIMALS

In dogs given 1.8 g/kg HES 150/0.70 for two hours after a standard burn of the hind quarters was inflicted, CVP and CO were decreased slightly after infusion (Lee and Clowes 1965; Lee *et al.* 1965). Increased in both TPR and coagulability observed after the burn were eliminated during the infusion of HES 150/0.70. This effect, however, waned by 4 hours after the infusion.

4.7.3.2. MAN

The resuscitation (as gauged by restoring haemodynamic indices) with HES 40/0.55 of patients experiencing burns has been favourable (Matsukawa 1976; Sato 1976).

4.7.4. ENDOTOXIN SHOCK

In dogs subjected to endotoxin shock and resuscitated 30 minutes later with 2 g/kg HES 450/0.70, haemodynamic characteristics were well maintained after infusion (Evangelista *et al.* 1969).

4.7.5. CARDIOPULMONARY BYPASS SURGERY

Various haemodilution agents are now used routinely to prime heart-lung machines for cardiac operations. Haemodilution has resulted in considerable conservation of blood, and diminution of plasma and corpuscle damage by decreasing the concentration of these elements in blood during extracorporeal circulation.

When HES 450/0.70 is used as the colloid component of an electrolyte solution (1 per cent concentration of HES 450/0.70), MAP, MAP per unit flow rate, and arterial PO_2 were adequately maintained during bypass surgery in patients (Lee *et al.* 1975). Plasma haemoglobin as well as the amount of blood given at operation were reduced when the HES 450/0.70 – electrolyte solution was used. Various combinations of HES 450/0.70 or HES 40/0.55 suspended in electrolyte solutions have been shown to be effective in experimental (Chu 1979; Kawashima 1972; Oishi *et al.* 1972) and clinical (Ckiba 1974; Kawasaki *et al.* 1971; Kubota 1975*a*; Matsui 1974; Mishler *et al.* 1975*b*; Miyata 1975; Nishijima 1976; Oishi *et al.* 1972; Sakauchi 1975; Yara 1974; Yoshitake *et al.* 1971) applications of extracorporeal circulation in which these solutions were used to prime heart-lung machines.

4.7.6. MISCELLANEOUS

Hydroxyethyl starch has also been shown to maintain haemodynamic and respiratory characteristics when used in the following surgical procedures: aneurysm correction (Iwatsuki 1977), reduction of cerebral ischaemia (Koch *et al.* 1978), chronic renal failure (Inoshita 1980), stabilization of cardiac infarcts (Yamashina 1975), during gynaecological operations (Iba 1975; Kitsutaka 1976; Nozue 1972; Suyama 1972), during laparotomy (Nishimoto 1976; Ura 1974), during neurosurgery (Kubota 1975*b*), during surgical procedures on older patients (Isikura 1976; Tagashira 1976), during obstetrical (Futani 1976; Hoshiai 1976; Maeda 1973) and orthopaedic (Maeda 1973) operations, during operations including spinal anaesthesia (Abe 1976; Kosaka 1975; Sakio 1976; Shiraki 1976), during general surgical procedures (*Lancet* 1971; Enomoto 1974; Fudeda 1976; Fujii 1976; Fujita *et al.* 1971; Goto 1971; Goto *et al.* 1971; Isa *et al.* 1971; Ishii *et al.* 1971; Ito 1972; Kanai 1975; Kaneko 1975; Kimura 1975, 1976; Kono *et al.* 1971; Kuba 1972; Kubota 1972; Matsuoka 1975, 1977; Miwa 1976; Miyazaki 1974; Muteki *et al.* 1971; Nakagawa 1973; Nishimura *et al.* 1972; Ogawa 1975; Oyama 1972; Satoshi *et al.* 1971; Shimizu *et al.* 1971; Sudo 1972; Sumida *et al.* 1971; Suzuki 1974; Takasugi 1976; Tanaka 1974; Toyama 1972; Tsushima *et al.* 1971; Yamada and Sakamoto 1975; Yamamoto 1972).

4.7.7. INFUSION THERAPY FOR PREVENTION OF ARTERIAL OR VENOUS THROMBOSIS

The prevention of thrombosis during and after surgery remains an important issue. The use of volaemic colloids to prevent intravascular coagulation may be related entirely to their haemodiluting action (Russell *et al.* 1966). Hydroxyethyl starch, however, has not proved to be effective in preventing arterial or venous clots in dogs or patients with damaged blood vessels (Arrants *et al.* 1969; Nokanishi 1976; Russell *et al.* 1966), even although hydroxyethyl starch decreases the formation of platelet aggregates in rabbits (as measured by the screen filtration pressure technique) if given in large quantities (0.4 or 1.2 g/kg) (Elmer *et al.* 1977).

5. The use of hydroxyethyl starch in blood banking

5.1. INTRODUCTION

During the past two decades, clinicians have not only depended on blood banks to provide whole blood but, with the use of more sophisticated chemotherapeutic regimens have needed specific blood components – namely granulocytes and platelets – to treat patients with thrombocytopenia and neutropenia. This use of the components of blood has been primarily due to the development of blood-cell separators, which are able to harvest large quantities of specific cell types from normal donors. In some circumstances, blood-cell separators have also been used to reduce the numbers of immature cells in the blood of patients with leukaemia (Vallejos *et al.* 1972, 1973). The efficiency of specific cell harvesting from normal donors and patients with blood-cell separators has been facilitated by the use of hydroxyethyl starch alone or in combination with glucocorticoid pretreatment in normal donors (Mishler 1975). In this chapter, the use of hydroxyethyl starch in the collection of several cell types using blood-cell separators will be discussed, and in addition to gravity leucapheresis, aspects of safety of this material in donors and effects on the function of harvested cells will be emphasized.

Blood banks are also concerned with the preservation of frozen cellular elements. Extracellular cryoprotectants such as hydroxyethyl starch, have been used in this context and some of this chapter will be devoted to the mechanism of reduced cellular injury in the presence of this material.

5.2. CENTRIFUGAL CYTAPHERESIS

5.2.1. LEUCAPHERESIS

5.2.1.1. EFFICACY OF HYDROXYETHYL STARCH WITH OR WITHOUT GLUCOCORTICOID PRETREATMENT

In a recent survey conducted by the American Bureau of Biologics of the Food and Drug Administration on the current use of leucapheresis and the collection of granulocyte concentrates in the United States, 68 per cent of the establishments responding indicated that when sedimenting agents were used during leucapheresis HES 450/0.70 was the agent preferred in 99 per cent of these centres (French 1980). The widespread use of HES 450/0.70 is probably based partly on historical considerations (McCredie and Freireich 1971) but, more importantly on its ability to significantly enhance the number of granulocytes collected from normal donors when added to the input-line of either the intermittent-flow centrifuge (IFC, Haemonetics Model 30 centrifuge) or continuous-flow centrifuge (CFC, Aminco-American Instrument Company; International Business Machines (IBM)) (Table 5.1). The overall number of lymphocytes harvested from normal donors using the Aminco blood-cell separator can also be increased by the addition of HES 450/0.70 to the input-line (Lenzhofer *et al.* 1978). In circumstances where a reduction in the peripheral blood concentration of immature cells in children or adult patients with acute or chronic myeloid leukaemia or acute lymphoblastic leukaemia would be advantageous, the addition of HES 450/0.70 to the input-line of either the IFC or CFC significantly increases collection efficiency (Graubner *et al.* 1978; Huestis *et al.* 1975*a*, 1976; Lane 1980; Vallejos *et al.* 1973; Wheeler *et al.* 1974; Woods *et al.* 1975).

TABLE 5.1 *The average number of granulocytes collected from normal donors by various blood-cell separators using only HES 450/0.70*

Centrifuge	Granulocytes collected ($\times 10^9$)		Reference
	No HES 450/0.70	HES 450/0.70	
Aminco	4.7	9.7	Mishler *et al.* (1974*a*)
IFC	3.4	10.2–13.3	Huestis *et al.* (1975*b*)
IFC	1.1–3.4	10.5–14.0	Huestis *et al.* (1975*c*)
IBM 2990 Aminco	2.0	10.0	McCredie *et al.* (1973)
IBM 2990 Aminco	1.8	9.1	McCredie *et al.* (1974)
IFC	–	7.1	Sussman *et al.* (1975)
IFC	–	7.6	Strauss *et al.* (1977)
Cytriage*	6.7	17.7	Ungerleider *et al.* (1977)
IBM 2990	2.7	10.1	
IFC	5.9	12.9	Bergmann *et al.* (1978)
IBM 2997	–	10.1	Hester *et al.* (1979)

* CFC, an automatic control system and a three-stage rotor

The survey conducted by the Bureau of Biologics also indicated that, of the establishments responding, 48 per cent used glucocorticoids to increase the blood level of granulocytes in normal donors before leucapheresis. In those centres using glucocorticoids, dexamethasone (DXM) and prednisone (PRED) accounted for 79 and 21 per cent of the drugs administered (French 1980). Whilst numerous studies have reported that glucocorticoid pretreatment of normal donors *alone* enhances the number of granulocytes collected by either IFC or CFC, still further numbers of cells can be harvested in circumstances where pretreatment with DXM (Table 5.2), PRED (Table 5.3), aetiocholanolone (AETIO) (Table 5.4), hydrocortisone (HYDRO, Table 5.5), or a combination of DXM and AETIO (Table 5.5) is combined with HES 450/0.70 to the input-line of either IFC or CFC cell separator.

The number of granulocytes collected by the various techniques depends on several factors: the number of litres of blood processed; the amount of HES

450/0.70 used per procedure; the type of blood-cell separator used; the pre-treatment schedule of glucocorticoid dosing; and the concentration of peripheral granulocytes in donor blood before collection.

TABLE 5.2 *The average number of granulocytes collected by various blood-cell separators using HES 450/0.70 combined with either single- or double-dose dexamethasone pretreatment of normal donors*

Treatment	Centrifuge	Granulocytes collected (× 10^9)		Reference
		Control	Experimental	
HES 450/0.70 + DXM*	IBM 2990	10.7	19.6	Higby *et al.* (1975)
				Mishler *et al.* (1975*a*)
	Aminco	–	13.0	Mishler *et al.* (1976)
	Aminco	–	24.0	Iacone *et al.* (1976)
	IFC	–	11.5	Schiffer *et al.* (1979)
	IFC	–	16.0–17.0	Huestis (1978)
	Aminco	2.9	21.5	Winton and Vogler (1978)
	IBM 2997	–	20.0	Hester *et al.* (1979)
HES 450/0.70 + DXM†	IBM 2990	10.7	20.3–25.5	Mishler *et al.* (1974*b*, 1975*a*)
	IBM 2997	–	19.8	Grindon and Coleman (1979)
	IFC	–	8.2	
	Aminco	2.9	32.2	Winton and Vogler (1978)

* Single-dose pretreatment
† Double-dose pretreatment

The number of litres of blood processed per procedure, using blood-cell separators, varies considerably from centre to centre and recent studies have proposed standards for efficient collection with either the IFC (Aisner *et al.* 1976) or CFC (Borberg 1978).

TABLE 5.3 *The average number of granulocytes collected by the Aminco blood-cell separator using HES 450/0.70 combined with pre-treating normal donors with a single dose of prednisone*

Treatment	Centrifuge	Granulocytes collected (× 10^9)		Reference
		Control	Experimental	
HES 450/0.70 + PRED	Aminco	–	25.6	Höcker *et al.* (1976)
	Aminco	5.6	14.4	Bearden *et al.* (1975, 1977)

Because of individual preferences in cell collection techniques, 15–60 g HES 450/0.70 may be given to normal donors or patients during a single procedure, even although the *average* dose is 30 g. In certain circumstances, normal donors may undergo three to four consecutive collection procedures. Because approximately 40–50 per cent of the administered dose of HES 450/0.70 remains in the blood after the initial collection procedure (Rock and Wise 1978, 1979;

TABLE 5.4 *The average number of granulocytes collected by CFC using HES 450/0.70 combined with pretreating normal donors with a single dose of aetiocholanolone*

Treatment	Centrifuge	Granulocytes collected ($\times 10^9$)		Reference
		Control	Experimental	
HES 450/0.70 + AETIO	IBM 2990 Aminco	2.0	22.0	McCredie *et al.* (1973)
	IBM 2990 Aminco	1.8	15.9	McCredie *et al.* (1974); Vallejos *et al.* (1975)
	IBM 2990 Aminco	–	22.1	McCredie and Freireich (1971)
	Aminco*	2.6	12.3	Hester *et al.* (1977)

* With disposable centrifuge bowl

Maguire *et al.* 1979*a*, 1981; Ring *et al.* 1980; Strauss and Koepke 1979, 1980), a dosage schedule has been proposed to alleviate the cumulative build-up of HES 450/0.70 in subsequent procedures (Mishler 1978*a*, *b*). The proposed schedule is: first procedure, 30 g; second, 18 g; and third procedure, 12 g. Although this sequence has not been verified, it may be appropriate for centres using normal donors on consecutive days.

TABLE 5.5 *The average number of granulocytes collected by CFC using HES 450/0.70 after pretreating normal donors with either a single dose of aetiocholanolone and dexamethasone or hydrocortisone*

Treatment	Centrifuge	Granulocytes collected ($\times 10^9$)		Reference
		Control	Experimental	
HES 450/0.70 + AETIO + DXM	IBM 2997	–	28.0	Hester *et al.* (1979)
HES 450/0.70 + HYDRO	Aminco	2.9	26.8	Winton and Vogler (1978)

The highest efficiency of granulocyte collection can be achieved in either of the various blood-cell separators when glucocorticoid pretreatment of normal donors is combined with the addition of HES 450/0.70 to the input-line. There presently exist several glucocorticoid pretreatment regimens for donors; but, most commonly, a *single* dose of either DXM or PRED is given orally 10–12 hours before collection or a *single* dose of either DXM or HYDRO is administered immediately or up to 3 hours before collection (Mishler 1977*b*, 1978*c*). In some centres, a *double*-dose DXM pretreatment regimen is used: oral dosing the night before combined with an intravenous injection several hours before collection. Depending on the dose of DXM, the number of litres processed and the type of blood-cell separator performing the procedure, the double-dose DXM regimen combined with HES 450/0.70 appears to result in the highest cell collection efficiency (see Table 5.2). Appropriate doses of glucocorticoids to be used for pretreatment purposes are presented in Table 5.6.

TABLE 5.6 *Glucocorticoids used in different centres for granulocyte collection, and their recommended dose for donor pretreatment**

Substance†	Mode of administration	Dosage (mg)
DXM	Oral, intravenous	6–10
AETIO	Intramuscular	6–9
HYDRO	Intravenous	100–400
PRED	Oral	40–60

* From Mishler (1977*b*)
† See text for abbreviations

In most circumstances, the concentration of granulocytes in the peripheral blood of donors before leucapheresis is a reasonable indicator of the yield of cells that will be harvested. Normally, the higher the pre-donation peripheral blood count, the greater the number of cells that can be collected. This relationship is depicted in Fig. 5.1 for cell collections from normal donors in which HES 450/0.70 or DXM pretreatment was not used. However, with the addition of HES 450/0.70 to the input-line, high cell yields can be obtained from normal donors with low pre-donation peripheral blood granulocyte levels. The use of HES 450/0.70, in combination with either *single* or *double* DXM donor pretreatment can also enhance the total number of cells harvested from normal donors with low pre-donation granulocyte counts.

One reason why the pre-donation peripheral blood granulocyte count is not an absolute indicator of eventual post-collection cell yield is the part played by the granulocytes stored in either the bone-marrow reserve or marginated granulocyte pool. Another reason is the degree of cell mobilization from one or both reserves during the collection procedure. Glucocorticoids release granulocytes stored in the bone-marrow reserve (Mishler and Williams 1980), whilst exercise or adrenaline release cells contained in the marginated granulocyte pool (Mishler 1977*b*). Most practitioners speculate that even in the absence of glucocorticoid pretreatment normal donors will mobilize cells from either of these reserves, accounting for the discrepancies between actual yields and those predicted theoretically.

The mechanism by which HES 450/0.70 significantly increases the yield of granulocytes collected with blood-cell separators from either normal donors or patients remains obscure. It was originally thought that the mechanism simply comprised decreasing the suspension stability (increased erythrocyte sedimentation rate (ESR)) of donor peripheral blood. This line of reasoning came from earlier observations that high cell yields were obtained from donors possessing an increased ESR. More recent data appear to indicate that hydroxyethyl starch may have a *specific* effect in certain separation devices (e.g. IBM 2997) (Hester, unpublished observations). This statement is supported by

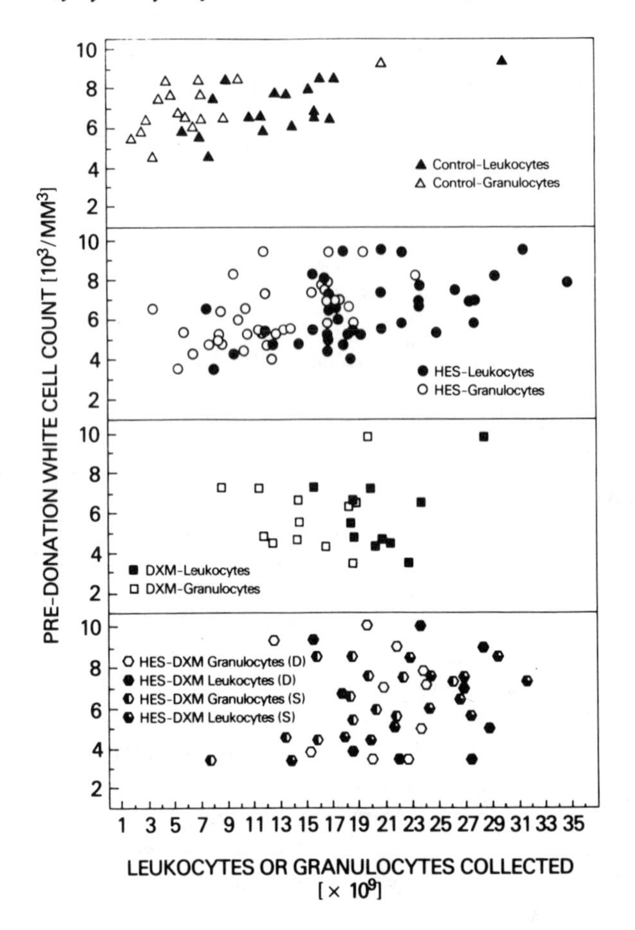

Fig. 5.1. Relationship between the pre-donation normal donor leucocyte count and post-leucapheresis cell collection yields. Ten litres of blood were processed at a flow rate of 40 ml/min. Some donors were pretreated with *single*-dose DXM only, whilst some donors received HES 450/0.70 in combination with *single*- (HES-DXM(S)) or *double*-dose (HES-DXM(D)) DXM pretreatment. (From Mishler 1975.)

observing that: (i) granulocyte collection efficiency is lowered considerably when anticoagulant is infused after the total dose of HES 450/0.70 has been administered, even though the donor still maintains an increased ESR (Hester, unpublished observations); (ii) granulocytes can be collected with the CFC using HES 264/0.43 (Hester, unpublished observations), even though HES 264/0.43 given in large doses (60 g) to normal donors had no effect on the ESR (Mishler *et al.* 1978*b*, 1979*b*); and (iii) infusion of HES 450/0.70 *before* commencement of the separation procedure, thus causing an immediate rise in the peripheral blood ESR, results in a collection efficiency similar to that in which the blood-cell separator was operated without HES 450/0.70 (Bergmann *et al.* 1978). However, in this same series of experiments, the infusion of HES 450/0.70 *during* the procedure resulted in a significant number of granulocytes collected.

If collection efficiency is related to a *specific* effect of hydroxyethyl starch in certain separation devices rather than to increases in the ESR, newer, more rapidly metabolized species, e.g. HES 264/0.43 (Strauss 1979) or HES 350/0.60 Mishler 1979*c*) may be used without compromising collection efficiency whilst decreasing the risk to the donor. Further clinical testing will be required to validate these conclusions.

Apheresis procedures using blood-cell separators in which hydroxyethyl starch has been added to the input-line have been employed for the procurement of eosinophils and for the treatment of diseases associated with eosinophilia (Gleich *et al.* 1979; Pineda *et al.* 1977).

5.2.1.2. EFFECTS ON CELLULAR FUNCTIONS

5.2.1.2.1. Granulocytes

After incubation in vitro

The suspension of granulocytes in plasma containing a *final* concentration of HES 450/0.70 ranging from 4–7 mg/ml does not significantly affect their ability to ingest either *Staphylococcus aureus* or latex particles, or their capacity to exclude trypan-blue dye (Mishler 1977*a*, 1978*a*; Roy *et al.* 1970, 1971). In addition, the morphology of these cells is normal. Under the same test conditions, the ability of granulocytes to ingest *S. aureus* in the presence of a *final* HES 450/0.70 concentration of 15 mg/ml is not compromised (Mishler 1977*a*, 1978*a*). Enhanced binding of *S. aureus* to granulocyte membranes concomitantly with increased rates of ingestion can be achieved *in vitro* by suspending cells in plasma containing HES 450/0.70, DXM and sodium-citrate (Mishler 1977*a*, 1978*a*, 1979*a*). It had been shown earlier (Ehlenberger and Nussenzweig 1977) that high concentrations of dextran 110 (20–60 mg/ml) were able to enhance ingestion by increasing the *binding* of particles to the granulocyte membrane. The exact manner in which this particular combination of all three agents (HES 450/0.70, DXM, and sodium-citrate) is better able to effect binding between *S. aureus* and the granulocyte membrane is unknown. The enhancing effect, however, is abolished when complement-depleted plasma is used as the suspending medium (Mishler 1977*a*).

Suspension of granulocytes in a solution of HES 450/0.70 and 0.9 per cent isotonic saline (Schiffer *et al.* 1975), or with plasma containing HES 450/0.70 (Mishler 1978*a*) had no significant effect on the bactericidal capacity of these cells against *S. aureus.* In these studies, the final concentration of HES 450/0.70 in the suspension media ranged from 7 to 15 mg/ml. Under similar test conditions, cells incubated in plasma containing HES 450/0.70, DXM, and sodium-citrate displayed normal bactericidal capacity and ultrastructure (Feliu *et al.* 1980; Mishler 1978*a*). A peripheral blood concentration of 6–7 mg/ml is normally achieved after an injection of 30 g HES 450/0.70 to healthy subjects (Mishler *et al.* 1977*a*; Rock and Wise 1978, 1979; Strauss and Koepke 1979, 1980).

TABLE 5.7 *Viability, morphology, and function of granulocytes in vitro when harvested by IFC, in which HES 450/0.70 was added to the input-line. In most instances, observations were made immediately, or 1–2 hours after collection*

	Morphology									
Viability	Light microscopy	Electron microscopy	Bactericidal activity	Chemotaxis	Random mobility	Adherence	Oxidative activity*	Phagocytosis	Fungicidal activity	Reference
Normal	Normal	Normal	–	Slightly increased	–	–	–	Normal	Normal	Glasser (1977)
Normal	–	–	Normal	Slighly increased	Normal	Normal	Normal	–	–	Strauss *et al.* (1977, 1979)
Normal	Normal	–	–	–	Normal	–	–	Normal	–	Schffer *et al.* (1975)
Normal†	–	–	Normal†	Normal†	–	–	–	Normal†	Normal†	Glasser and Huestis (1979)
Normal†	–	–	Normal†	Normal†	–	–	–	Normal†	Normal†	Glasser *et al.* (1977)

* Superoxide anion production, NBT reduction, hexose monophosphate shunt activity, and chemiluminescence in resting and phagocytic cells
† Donors were pretreated with DXM

After harvesting by centrifugal leucapheresis

Intermittent-flow centrifugation The functional capacity *in vitro* of granulocytes collected from normal donors (with or without pretreatment with DXM) using the IFC in which a solution of HES 450/0.70 and sodium-citrate was added to the input-line was not impaired when assessed immediately, or 1–2 hours after harvesting (Table 5.7). The effects of glucocorticoids on neutrophil function have recently been reviewed (Mishler 1977*b*) and the results have shown no impairment when these agents are used at pharmacological doses.

The survival *in vivo* of granulocytes harvested by IFC in which a solution of HES 450/0.70 and sodium-citrate was added to the input-line has been assessed recently (Price and Dale 1978). Collected granulocytes were labelled with [^{32}P]-diisopropylfluorophosphate (DF^{32}P) and injected after harvesting. When IFC cells were not stored after collection, the 1-hour recovery was 27.6 per cent with a $T_{1/2}$ of 4.1 hours; labelled cells obtained by phlebotomy had a 1-hour recovery of 33.1 per cent and a $T_{1/2}$ of 5.8 hours.

Continuous-flow centrifugation The functional capacity *in vitro* of granulocytes harvested from normal donors (with or without pretreatment with steroids) using the CFC, in which HES 450/0.70 was added to the input-line was not impaired when assessed immediately, or 1–2 hours after collection (Table 5.8). Additional studies have also indicated that function *in vitro* of granulocytes collected with HES 450/0.70, in which donors were pretreated with DXM were not functionally impaired when cells were harvested by CFC (Iacone *et al.* 1976).

The survival *in vivo* of granulocytes collected by CFC in which HES 450/0.70 was added to the input-line has been assessed (McCullough *et al.* 1976). Collected cells were labelled with DF^{32}P and injected after harvesting. The post-transfusion recovery of cells collected by ordinary phlebotomy averaged 52 per cent compared with 34 per cent for granulocyte concentrates prepared by CFC. The $T_{1/2}$ was 3.8 hours for cells collected by phlebotomy and 3.0 hours for CFC-prepared granulocytes.

5.2.1.2.2. Monocytes

The function of monocytes harvested by IFC, in which HES 450/0.70 was added to the input-line has been studied recently (Eastlund *et al.* 1979, 1980). Monocyte chemotaxis and migration were not significantly altered during IFC and exposure to HES 450/0.70.

5.2.1.2.3. Lymphocytes

Intermittent-flow centrifuge

Total B-cell and T-cell populations together with lymphocyte reactivity to mixed leucocyte culture, phytohaemagglutinin (PHA), pokeweed mitogen and concanavalin-A were determined in normal donors exposed to HES 450/0.70 during cytapheresis performed with the IFC (Braine *et al.* 1980). The results

TABLE 5.8 *Morphology (by light microscopy) and function in vitro of granulocytes harvested by CFC in which HES 450/0.70 was added to the input-line. In most instances, observations were made immediately, or 1–2 hours after collection*

Morphology	Bactericidal activity	Chemotaxis	Oxygen consumption	Phagocytosis	NBT reduction	Reference
–	Normal	–	–	–	Normal	Mishler *et al.* (1974*a*)
–	Normal*	Normal*	–	–	–	Ungerleider *et al.* (1977)
Normal†	–	–	–	Normal†	–	Hester *et al.* (1979)
Normal	Normal	Normal	Normal	–	Normal	McCullough *et al.* (1976)

* Cytriage (see Table 5.1)
† Donors were pretreated with steroids

indicated that lymphocytes harvested from these donors did not differ from cells obtained from the peripheral blood of controls. Lymphocytes collected from normal donors undergoing IFC in which HES 450/0.70 is added to the input-line generated a neutrophil chemoattractant in a dose- and time-dependent fashion after appropriate stimulation by antigens or PHA (Eastlund *et al.* 1979, 1980.

Continuous-flow centrifugation

Lymphocytes harvested by CFC from AETIO and/or DXM pretreated normal donors incorporated normal amounts of labelled thymidine when cultured with PHA and streptolysin-O (Hester *et al.* 1979).

5.2.1.3. EVALUATION OF DONOR SAFETY

5.2.1.3.1. Tests of organ function

The effect of hydroxyethyl starch on hepatic and renal function has been discussed in Section 3.9. Tests of organ function specifically performed on normal donors undergoing cytapheresis will be presented in this section.

Normal donors (with or without glucocorticoid pretreatment) undergoing *single* or *multiple* granulocyte (with or without platelets) collection procedures with either the IFC or CFC, in which HES 450/0.70 was added to the input-line (15-45 g per procedure), did not experience abnormalities in tests of hepatic or renal function (Hester *et al.* 1975, 1979; Maguire *et al.* 1979*a*, 1981; McCredie *et al.* 1974; Mishler 1975; Mishler *et al.* 1974*a*, 1975*a*, 1976; Strauss and Koepke 1979, 1980; Sussman *et al.* 1975). Long-term follow-ups on normal donors who underwent cytapheresis one or more years previously revealed no toxic effects of the administered HES 450/0.70 (Maguire *et al.* 1979*a*) or HES 450/0.70 and AETIO (McCredie *et al.* 1974).

5.2.1.3.2. Effects on constituents of blood and plasma

The effect of hydroxyethyl starch on constituents of blood and plasma has been previously discussed in Section 4.3. As pointed out in Chapter 4, the extent of changes in the concentration of various constituents in blood or plasma depends on the amount of hydroxyethyl starch infused, the water-binding potential, the intravascular persistence, and whether whole blood or plasma was withdrawn before or during the infusion. Normally, 15-45 g HES 450/0.70 are infused during the collection process of cytapheresis. The water-binding potential of HES 450/0.70 is approximately 20 ml water/g intravascular colloid (see Table 4.7). The intravascular persistence of HES 450/0.70 is prolonged because of the high MS, with approximately 40-50 per cent of the administered dose remaining 24 hours after the initial infusion (Mishler *et al.* 1977*a*; Maguire *et al.* 1979*a*, 1981; Rock and Wise 1979. Strauss and Koepke 1979, 1980). In normal donors undergoing multiple collection procedures, HES 450/0.70 will accumulate in the intravascular space during subsequent donations (Maguire *et al.* 1979*a*, 1981; Strauss and Koepke 1979, 1980). During cytapheresis, small quantities of erythrocytes and plasma are removed concomitantly with larger amounts of leucocytes and platelets. These factors should be remembered when reviewing the changes in the constituents of blood and plasma during cytapheresis.

In normal donors (with or without glucocorticoid pretreatment) undergoing centrifugal cytapheresis in which HES 450/0.70 is added to the input-line the low haemoglobin, haematocrit, albumin, and total protein levels during the *initial* collection procedure return to approximate baseline values 24 hours later (Hester *et al.* 1975, 1979; Mishler 1975; Mishler *et al.* 1974*a*, *b*, 1975*a*, 1976; Sussman *et al.* 1975). In donors undergoing multiple procedures, the fall in haemoglobin, haematocrit, albumin, and total proteins will depend on several factors: the time interval between consecutive procedures; the amount of HES 450/0.70 administered per procedure; the amount of erythrocytes and plasma removed during each donation. Generally speaking, a shorter time interval between consecutive procedures, combined with moderate (30 g) to high (45 g) doses of HES 450/0.70 will result in noticeable declines in blood constituents although these changes are temporary. In addition, the removal of 30–50 ml of packed erythrocytes per procedure will further decrease haemoglobin and haematocrit values. In attempting to avoid these changes occurring in donors undergoing multiple procedures, lesser amounts of HES 450/0.70 could be used on subsequent donations (Mishler 1978*b*), although this regimen has not been evaluated clinically. The reader is advised to review the work of several authors (Freireich *et al.* 1978; Hester *et al.* 1975; Koepke *et al.* 1981; Mishler 1975; Mishler *et al.* 1974*a*; Strauss *et al.* 1980*a*,*b*) for changes taking place in the blood or plasma constituents of normal donors undergoing multiple procedures.

Patients with leukaemia undergoing repeated leucapheresis for up to 26 months tolerated HES 450/0.70 well without signs of morbidity, except that anaemia occurred as a result of an intensive collection regimens (Vallejos *et al.* 1973), probably as the result of removal of reticulocytes (Woods *et al.* 1975).

5.2.1.3.3. Changes in the plasma volume

The effect of the infusion of hydroxyethyl starch on changes in body fluid compartments is discussed in Section 4.6. In normal circumstances, HES 450/0.70 will increase the plasma volume in direct proportion to the amount of solution infused. In centres using normal donors on multiple occasions, Rock and Wise (1978, 1979) have measured the changes in plasma volume of donors who received daily doses of 30 g HES 450/0.70 for four consecutive days. All donors showed marked and progressive increases in plasma volume; the average increase was 40 per cent, with one subject increasing his plasma volume by 1 litre. In only one subject had values returned to baseline by the sixth day; in the others, values were still increased by at least 500 ml. The expanded plasma volumes cited in this study are due to the persistence of HES 450/0.70 in the intravascular space and to its high water-binding capacity. Recent studies have recommended that newer, more rapidly-metabolized species of hydroxyethyl starch, e.g. HES 264/0.43 (Strauss 1979) or HES 350/0.60 (Mishler 1979*c*) are clinically evaluated with the hope of reducing the large changes in the plasma volume of donors undergoing multiple collection procedures.

5.2.1.3.4. Effects of coagulation

The effects of hydroxyethyl starch on coagulation, fibrinolysis, and haemostasis have been discussed in Section 3.8. Based on studies in animals and in man, marked alterations in coagulation are observed when over 90 g hydroxyethyl starch are given as a *single* bolus infusion.

During cytapheresis, in addition to the fact that donors are given HES 450/0.70, the processed blood is exposed to the foreign surfaces of the apheresis apparatus, and platelets and granulocytes, both of which contain thromboplastic substances, are concentrated. This exposure of plasma and cells to foreign surfaces in high concentration could result in activation of clotting, thrombus formation, and the potential for return of thrombogenic substances to the donor.

In normal donors undergoing a *single* cytapheresis procedure to which 15–45 g HES 450/0.70 were added to the input-line, results of coagulation tests (PT, PTT, ACT, and bleeding time) usually gave results within normal limits (Hester *et al.* 1975, 1979; Maguire *et al.* 1978, 1979*c*; McCredie *et al.* 1974; Mishler 1975; Mishler *et al.* 1974*a*, *b*, 1975*a*, 1976; Sussman *et al.* 1975) after termination of cell donation, even although abnormalities in PT and PTT have been reported (Kisker *et al.* 1979; Strauss 1981; Strauss and Koepke 1979, 1980). However, these abnormalities, along with significant reductions in the activities of Factors V, VIII, and fibrinogen, were slight. These abnormalities were not considered to be of sufficient magnitude to implicate excessive consumption of clotting factors, as might happen if activation of the clotting cascade occurred during apheresis, or to indicate the return of thrombogenic substances to the donor.

Slight prolongations of PT and PTT were observed in normal donors undergoing multiple cell collection procedures to which 15–30 g HES 450/0.70 were added to the input-line (Mishler *et al.* 1974*a*, 1976). In donors given 30 g HES 450/0.70 on each of four consecutive days (total 120 g), there were prolongations of the PTT while the PT remained within normal limits (Rock and Wise 1979).

Massive extracorporeal blood clotting has been reported during IFC in which a mixture of HES 450/0.70 and sodium-citrate was added to the input-line (Drescher *et al.* 1978). Further investigation of this episode revealed that during the cell collection procedure, *excess* citrate was infused initially, followed by *inadequate* amounts delivered at the end of the donation. The reason was due to a marked and inapparent sodium-citrate concentration gradient occurring due to the significant difference in the specific gravity of the two solutions. It was recommended that the admixture of HES 450/0.70 and sodium-citrate should be forceably agitated for 30 seconds before starting the collection procedure.

5.2.1.3.5. Changes in the complement system

During the collection of granulocytes from normal donors with IFC, in which

HES 450/0.70 was added to the input-line, alterations in serum complement proteins were noted (Strauss *et al.* 1978, 1980*c*). Significant decreases in C1, C2, C3–C9, and factor B occurred in donor blood after 60 minutes of apheresis. However, all deficiencies except B corrected spontaneously as the procedure continued. In contrast, concentrations of C2, C4, C3–C9, CH50, and factor B remained decreased in machine efferent fluids throughout the procedure. Similar, although not identical, changes accompanied platelet collections by IFC performed without HES 450/0.70 (Strauss and Koepke 1979, 1980). The mechanisms affecting these changes are unknown, but it seems clear that alterations of many complement proteins accompany apheresis procedures whether or not HES 450/0.70 is used (Strauss and Koepke 1979). Complement activation, however, seems a less likely explanation. Instead, complement proteins may be lost by adsorption on to the surfaces of the IFC software (Strauss *et al.* 1980*c*).

5.2.1.3.6. Side effects on donors

Several studies have reported no major untoward effects in normal donors receiving either AETIO or AETIO plus DXM during *single* or *multiple* collection procedures to which HES 450/0.70 was added to the input-line of the CFC (Hester *et al.* 1975, 1979; McCredie *et al.* 1974). Chills, fever, mild influenza-like symptoms, mild headaches, muscle pains and mild peripheral oedema (of the lower extremities) were reported. Some donors dosed with AETIO manifested a 1–2 °C increase in temperature. Some donors given HES 450/0.70 experienced symptoms associated with plasma volume expansion. These symptoms were restricted to mild headaches and oedema and were difficult to distinguish from those elicited by AETIO. Normally no medication was required, or only the use of a mild analgesic for 24 hours or less. No allergic or anaphylactic reactions were observed in over 150 donors, even after repeated use of HES 450/0.70 in the same donor (see Table 3.5 for the number of donors undergoing multiple procedures with HES 450/0.70 and glucocorticoids).

Additional studies have shown that HES 450/0.70 and DXM were well tolerated by normal donors except in 5 out of twenty-eight instances, where chills, fever, and headache were observed (Mishler *et al.* 1974*b*, 1975*a*). Chills were thought to be associated with lowering of the body temperature by the extracorporeal circuit. In other studies, no untoward effects were reported during eighty-three cell collection procedures (Mishler 1975; Mishler *et al.* 1974*a*, 1976). Additional studies have reported circumoral paraesthesias, increased neuromuscular excitability, and syncopal symptoms, but these are believed to be due to the infusion of sodium-citrate (Huestis *et al.* 1975*c*; Sussman *et al.* 1975).

Recently, a normal male volunteer developed lichen planus in association with leucapheresis in which HES 450/0.70 was used to increase the yield of granulocytes (Bode and Deisseroth 1981). Although there was no direct proof of a causative relationship, these authors, nevertheless, implied that the temporal association of the onset of initial symptoms and HES 450/0.70 infusion strongly supported such a conclusion.

TABLE 5.9 *The average number of platelets harvested by IFC from normal donors (with or without glucocorticoid pretreatment) using HES 450/0.70*

Treatment	Centrifuge	Platelets collected ($\times 10^{11}$)	Reference
HES 450/0.70	IFC	4.5	Strauss *et al.* (1977)
	IFC	3.1	Strauss *et al.* (1979)
	IFC	6.7	Huestis (1978)
	IFC	7.7	Sussman *et al.* (1975)
HES 450/0.70 + DXM	IFC	High*†	Schiffer *et al.* (1979)
	IFC	7.4†	Huestis (1978)
		7.6*	

* Single intravenous dose
† Single dose

5.2.2. COMBINED LEUCAPHERESIS AND PLATELETPHERESIS

Centrifugal leucapheresis in the presence of HES 450/0.70 is being increasingly used to provide a combined platelet and granulocyte preparation for transfusion into patients suffering from both thrombocytopenia and neutropenia (Katz *et al.* 1978; Kisker *et al.* 1979; Maguire *et al.* 1979*c*; Rock *et al.* 1980; Schiffer *et al.* 1979; Wong and Rock 1979). In these circumstances, neither the function of harvested granulocytes nor of platelets should be compromised by the presence of HES 450/0.70 during the collection period. The function *in vitro* of granulocytes exposed to HES 450/0.70 and glucocorticoids has been discussed. The effectiveness of using HES 450/0.70 (with or without DXM donor pretreatment) during the harvesting of a combined platelet–granulocyte preparation, and the function *in vitro* of the platelets contained in the concentrate will now be reviewed.

5.2.2.1. EFFICACY OF HYDROXYETHYL STARCH

The addition of HES 450/0.70 to the input-line of the IFC during combined platelet–granulocyte harvesting has been reported to result in high platelet yields (Table 5.9). The pretreatment of donors with DXM and the addition of HES 450/0.70 has also been reported (Table 5.9), with some suggestion that higher yields of platelets are collected with this technique (Huestis 1978). The number of platelets collected using HES 450/0.70 alone or in combination with DXM will depend on several factors, including the quantity of blood processed and the level of platelets in the peripheral blood of donors before donation. Considering all factors of the collection procedure, the use of HES 450/0.70 appears to be an efficient means of increasing the total number of platelets harvested by IFC.

5.2.2.2. EFFECT ON CELLULAR FUNCTION

After incubation ***in vitro***

Platelets prepared by the technique of manual apheresis were incubated for 2–4

TABLE 5.10 *Morphology, ultrastructure, and function of platelets in vitro when prepared by either IFC or CFC in which HES 450/0.70 was added to the input-line. In most instances, observations were made immediately or 1–2 hours after collection*

	IFC					CFC
	1, 2	3	4	5	6	7
Morphology	Normal	–	Normal*	–	–	–
Ultrastructure	Normal	–	–	Normal†	–	–
Adhesiveness	–	Normal	–	–	Normal	Normal
Osmotic recovery	–	–	Normal*	–	–	–
Prostaglandin synthesis	Increased	–	–	–	–	–
Aggregation:						
Spontaneous	–	Not detected	–	–	Not detected	–
ADP	Normal	Normal	–	Normal	Normal	Normal
Ephinephrine	Normal	Normal	–	Decreased	Normal	–
Thrombin	Normal	–	–	–	–	–
Collagen	Normal	Normal	–	Normal	Normal	Normal
Ristocetin	Normal	–	–	–	–	–
[^{14}C]-Serotonin release:						
Collagen	Increased	–	–	–	–	–
Thrombin	Increased	–	–	–	–	–
ADP	Normal	–	–	–	–	–
Ristocetin	Normal	–	–	–	–	–
Epinephrine	Normal	–	–	–	–	–
ATP-release:						
Collagen	Slight decrease	–	–	–	–	–
Thrombin	Slight decrease	–	–	–	–	–
ADP	Slight decrease	–	–	–	–	–
Ristocetin	Slight decrease	–	–	–	–	–
Epinephrine	Slight decrease	–	–	–	–	–
Size-distribution curve:	Normal	–	–	–	–	–

* Stored at 22 °C for 48 hours
† Except for a decrease in glycogen granule content
1 – Wong and Rock (1979); 2 – Rock *et al.* (1980); 3 – Maguire *et al.* (1978); 4 – Katz *et al.* (1978); 5 – Maguire *et al.* (1979*c*, 1980*b*); 6 – Maguire *et al.* (1979*b*, 1980*a*); 7 – Farrales *et al.* (1975, 1977)

hours (at 25 °C) with either HES 450/0.70 or 0.9 per cent isotonic saline (Schiffer *et al.* 1975). After 2 hours, the release of nucleotides or of serotonin from HES 450/0.70-incubated platelets was not significantly affected. No change in platelet morphology occurred after exposure to HES 450/0.70 for up to 4 hours. Of these cells, over 80 per cent remained disc-like and the appearance of dendritic forms was rare. No alteration in platelet size distribution, in the calculated mean platelet volume, or in the availability of platelet factor-3, occurred after 4 hours of incubation with HES 450/0.70. The ability of platelets to aggregate when incubated with ADP was also not affected.

After harvesting by cytapheresis

The morphology, ultrastructure, and function *in vitro* of platelets prepared by a combined platelet-granulocyte apheresis technique appear to be normal (Table 5.10). Prostaglandin synthesis, the release of ATP and serotonin in platelets exposed to collagen, thrombin, ADP, epinephrine, or ristocetin and aggregation, adhesiveness, and size-distribution are not compromised when HES 450/0.70 is used in this type of cell collection. In some centres using DXM to pretreat donors before the combined platelet-granulocyte collection procedure (Huestis 1978; Schiffer *et al.* 1979), DXM *alone* does not interfere with normal platelet function *in vitro* (Lichtenfeld and Schiffer 1979).

5.3. GRAVITY LEUCAPHERESIS

Leucaphereses with IFC or CFC are relatively complicated procedures using costly mechanical devices and requiring highly skilled personnel. The technique of filtration leucapheresis (FL) is simpler and more efficient but subtle changes result from the contact of granulocytes with the nylon fibres (Djerassi *et al.* 1977). With the object of permitting widespread and routine harvesting of transfusable granulocytes, a new technique has been developed for collecting granulocytes by *gravity* alone, unassisted by centrifugation but with the aid of HES 450/0.70 (Djerassi 1977). This method compares favourably with either IFC, CFC, or FL in terms of the yield of granulocytes per unit of blood (Aisner *et al.* 1977, 1979) and the results can be further improved by the administration of glucocorticoids to donors before the start of the collection procedure (Djerassi 1977). Granulocytes contained in the final concentrate prepared by gravity leucapheresis using HES 450/0.70 possess normal morphology as well as normal function *in vitro* (phagocytic index and NBT reduction) (Aisner *et al.* 1977; Poon and Wilson 1980). The original technique of Djerassi (1977) is described in Appendix 4; modifications of the original procedure have been reported by Aisner *et al.* (1977, 1979) and by Poon and Wilson (1980).

5.4. PREPARATION OF LEUCOCYTE-POOR BLOOD

Transfusion of leucocyte-poor red blood cell (LP-RBC) suspensions may aid in the management of patients experiencing recurrent severe febrile reactions secondary to sensitization to infused donor leucocytes. In this regard, LP-RBC suspensions may possibly serve a preventive value in minimizing recipient sensitization to leucocyte-specific and histocompatibility antigens carried on donor white blood cells. Possible consequences of such sensitization would include: (i) liability to severe febrile reactions after future whole-blood transfusions (Brittingham and Chaplin 1957); (ii) early rejection of homologous donor grafts (Pierce *et al.* 1971); (iii) predisposition to alloimmune neonatal neutropenia in subsequent pregnancies (Walford 1969); (iv) aggravation of pre-existing severe leucopenia after reactions to donor leucocytes in platelet concentrates administered to myelosuppressed patients (Herzig *et al.* 1974).

Various techniques have been used for the preparation of LP-RBC suspensions (Polesky *et al.* 1973). A simple technique using the sedimentation of whole blood or packed red blood cells (RBCs) in HES 450/0.70 has recently been described (Dorner *et al.* 1974, 1975), which on average removes over 93 per cent of the leucocytes present in the original donor unit. The technique is described in Appendix 5.

The effect of the preparation of LP-RBC with HES 450/0.70 on cellular function and preliminary clinical results are described below.

5.4.1. FUNCTION *IN VITRO*

Measurements of erythrocyte 2,3-DPG and oxyhaemoglobin dissociation characteristics (p. 50) after *two* sedimentations (through step 7, Appendix 5) of CPD donor blood stored at 2–4 °C for 3–7 days were similar (not statistically or biologically significantly different) to control values obtained immediately before donor unit processing.

5.4.2. FUNCTION *IN VIVO*

Aliquots of blood from six donor units labelled with $Na_2[^{51}Cr]O_4$ and subjected to the standard sedimentation procedure (through step 7, Appendix 5) were re-infused into the original donor. The $[^{51}Cr]$-erythrocyte survivals (24 hours after injection) were in the normal range of 72-hour-stored CPD blood.

5.4.3. CLINICAL RESULTS

Two hundred and ninety-six units of LP-RBCs prepared by the HES 450/0.70-sedimentation technique were transfused to seventy patients. In most cases (294 units), the LP-RBCs were transfused without incident. One mild and one moderate febrile reaction were observed but appeared not to be related to the infusion of the HES 450/0.70 (80–120 ml of 3 per cent (w/v) HES 450/0.70 per unit of LP-RBCs).

5.5. FREEZING OF CELLULAR ELEMENTS

Simple freezing and thawing of cells or tissues results in cell dissolution and in most circumstances is lethal (Mitchell 1979). The mechanisms of freeze injury are known to include simple thermal shock, intracellular and extracellular formation of ice with removal of water, cell shrinkage, sodium choride (NaCl) lysis, and production of high concentrations of solute (Scheiwe *et al.* 1979). Denaturation injury to essential cellular elements including dehydration of proteins and lipoprotein membrane modifications associated with abnormal disulphide bonding has also been described. Methods of avoiding or preventing some or all of these adverse effects have concentrated on variation of the rate of cooling, the design of suitable containers with desirable thermodynamic properties to induce rapid transfer of heat, and the use of organic molecules added to the freezing admixture to alter the degree of hydration and limit losses due to formation of ice.

The ultimate purpose of any cryoprotectant, whether it penetrates (intracellular) the cell or not (extracellular), is the elimination of freeze injury (McGann 1978). If freeze injury could be prevented, the ability to freeze-preserve various cellular elements for longer periods than are currently available using liquid-temperature (4 °C or 22 °C) storage would have tremendous implications. Currently, intracellular agents – namely dimethylsulphoxide (Me_2SO) or glycerol – are used when RBCs, platelets, or granulocytes are stored at low temperatures (−80° to −196 °C). These agents are effective in preventing freeze injury; however, they must be removed by sophisticated washing procedures, even though Meryman and Hornblower (1978) have developed a new method to deglycerolize RBCs with hydroxyethyl starch sedimentation.

An ideal cryoprotectant, however, would be one that could be transfused directly with the thawed cells without prior washing, thus eliminating the need for futher processing. During the last decade, numerous investigators (Allen *et al.* 1976, 1977, 1978; Baar 1973; Forest *et al.* 1972; Knorpp *et al.* 1967*a*, *b*, 1968, 1969, 1971*a*, *b*; Lionetti and Hunt 1974, 1975; Lionetti *et al.* 1975*a*, 1976; Lionetti and Callahan 1979; Robson 1970; Weatherbee *et al.* 1972, 1974*a*, *b*, 1975*a*, *b*, 1979) have developed techniques by which RBCs can be frozen in the presence of the extracellular cryoprotectant, hydroxyethyl starch. The main objective of these investigations was the development of a freeze–thaw technique that would allow the direct transfusion of the RBC–hydroxyethyl starch admixture with minimal post-thaw manipulation. The results of these studies are summarized below. Bendavid and Gavendo (1971, 1972) have also described a technique by which RBCs are frozen in the presence of both hydroxyethyl starch and polyvinylpyrrolidone. Other investigators have developed techniques in which granulocytes, platelets, and lymphocytes can be preserved in the frozen state with hydroxyethyl starch alone or in combination with Me_2SO. These results will also be summarized in subsequent sections.

5.5.1. ERYTHROCYTES

The freezing of full-units of RBCs with hydroxyethyl starch includes many factors: design of a species of hydroxyethyl starch with a suitable $M_{\bar{w}}$ and MS; development of optimal rates of freezing and thawing to preserve cellular function, and techniques to assess the effect of hydroxyethyl starch in reducing freeze injury. These various factors will be discussed below.

5.5.1.1. DESIGN OF A SUITABLE CRYOPROTECTANT

The design of an appropriate species of hydroxyethyl starch for use during the freeze-preservation of human RBCs at low temperatures includes not only physicochemical considerations but may also include practical and clinical aspects as well (Mishler 1979*b*).

Physicochemical properties

Greenwood *et al.* (1975, 1977) found that when human RBCs were frozen in small (50 ml) aliquots, the molecular size of hydroxyethyl starch had an important effect on cell recovery after thawing. These investigators noted that consistently *higher* recoveries of RBCs were obtained when *low*-viscosity hydroxyethyl starch was used (Table 5.11), and that within the viscosity range studied, the post-thaw recovery of cells was inversely proportional to the viscosity of hydroxyethyl starch (Greenwood *et al.* 1977). Lionetti *et al.* (1976) reported that low-viscosity hydroxyethyl starch yielded greater post-thaw recoveries than high-viscosity hydroxyethyl starch when RBCs were frozen in full- (>400 ml) units. Unlike dextran, osmotic effects are much less for hydroxyethyl starch of comparable viscosity potential (see Section 1.4.4.3).

TABLE 5.11 *Variation of the viscosity of hydroxyethyl starch and the cryoprotective effect**

Study	High-viscosity†		Low-viscosity‡	
	Recovery (%)	Saline stability (%)	Recovery (%)	Saline stability (%)
A	95	86	98	84
B	95	87	97	89
C	96	87	96	86
D	95	88	96	89
E	92	76	94	89
F	91	79	92	80
Mean	94	84	96	86

* From Greenwood *et al.* (1977)
† Commercial hydroxyethyl starch, viscosity 180 cp for 30 per cent solution at 20 °C
‡ Commercial hydroxyethyl starch, viscosity 90 cp for 30 per cent solution at 20 °C

Greenwood *et al.* (1975, 1977) also compared the effect of two levels of molar hydroxyethyl group substitution (MS = 0.39 and 0.80) on the post-thaw recovery and saline stability of human RBCs. Molar substitution was shown not to influence the recovery of intact cells but had a significant effect on the stability of cells after thawing (Fig. 5.2). At thè *higher* level of molar substitution, saline stability was increased except at high concentrations of hydroxyethyl starch. The differences between tests of recovery and saline stability will be discussed later (p. 142). A similar but less pronounced effect was observed in freezing mixtures to which plasma had been added. However, in later tests these investigators observed no significant difference between otherwise similar samples of hydroxyethyl starch differing in molar substitution between 0.70 and 0.80.

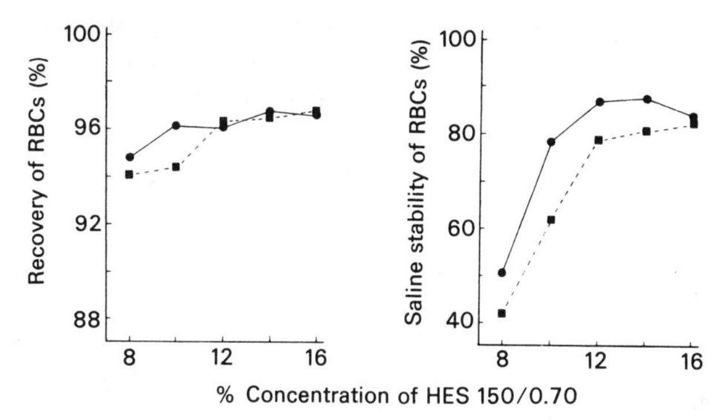

Fig. 5.2. Post-thaw recovery and saline stability of human RBCs protected during freezing with laboratory-prepared hydroxyethyl starch at levels of molar substitution of 0.39 (■) and 0.80 (●). Plasma was not added to the admixture of RBCs and hydroxyethyl starch before freezing. (From Greenwood *et al.* 1977.)

Practical and clinical consideration

Practical considerations may include the dilemma between the transfusion of a viscous admixture of RBCs and hydroxyethyl starch and the desire to have the highest concentration of hydroxyethyl starch yielding superior cryoprotective effects (Lionetti *et al.* 1976). This problem may, however, be partially solved by means of a simple post-thaw washing procedure (Lionetti and Callahan 1979; Weatherbee *et al.* 1979) (see Appendix 6).

Clinical aspects include designing a species of hydroxyethyl starch having the highest MS which after transfusion will allow the infused material to be catabolized and excreted within a reasonable period of time. Preliminary studies both in monkeys (Weatherbee *et al.* 1974*c*) and humans (Mishler and Parry 1979) in which RBCs and large quantities of the HES 150/0.70 cryoprotectant were *directly* transfused after thawing have shown that the HES 150/0.70 was removed from blood within a reasonable period. The infusion of the RBC–hydroxyethyl starch admixture was well tolerated without evidence of bleeding or bleeding tendencies. Serum chemistry values were not significantly altered either in monkeys or humans after transfusion.

The use of the high molar substituted (0.70), low-viscosity HES 150/0.70 in these preliminary clinical trials appears to satisfy both physicochemical and clinical considerations for an ideal extracellular cryoprotectant. Further testing of other species of hydroxyethyl starch will be required to substantiate this claim.

5.5.1.2. THE OPTIMAL CONCENTRATION OF HYDROXYETHYL STARCH IN THE CRYOPROTECTANT SOLUTION

Theoretical

Differential thermal analysis has been used to determine the optimal concentration of HES 450/0.70 required to reduce the concentration of NaCl below a certain level during freezing of RBCs *without* plasma (Körber and Scheiwe 1977, 1980; Scheiwe *et al.* 1979). This value for HES 450/0.70 was based on the assumption that the overall concentration of NaCl was the only injurious variable, and the ternary system (HES 450/0.70, water, and NaCl) was in equilibrium. Assuming that an overall concentration of NaCl in the region of 4.4 per cent is lethal, then a final concentration of 12 per cent HES 450/0.70 in the cryoprotectant medium can effectively hinder concentration of NaCl beyond this level of 4.4 per cent during freezing.

Practical small-unit (25–55 ml) freezing

Adopting the saline stability test to assess the viability of human RBCs after freezing, Greenwood *et al.* (1977) reported that when plasma was not added to the cryoprotectant medium, a 16 per cent concentration of low-viscosity hydroxyethyl starch yielded a value of 82 per cent. Saline stability values of 83–85 per cent have been reported when human RBCs were frozen without plasma in a medium containing either 14 per cent HES 150/0.70 (Weatherbee *et al.* 1974*b*, 1975*a*) or 15 per cent HES 450/0.70 (Knorpp *et al.* 1967*a*,*b*). Human RBCs frozen in a medium containing 10 per cent HES 150/0.70 without plasma were not as viable as cells preserved with 14 per cent HES 150/0.70 (Weatherbee *et al.* 1975*b*).

Controversy exists over the value of adding plasma to the cryoprotectant medium containing hydroxyethyl starch. Greenwood *et al.* (1977) and Lionetti and Hunt (1974, 1975) have reported that the post-thaw saline stability was approximately 87 per cent when human RBCs were processed in a medium containing 20 per cent plasma and a concentration of low-viscosity hydroxyethyl starch of 12–16 per cent. Saline stability is also maintained in human RBCs preserved in a medium of 20 per cent albumin and 12 per cent HES 150/0.70 (Baar 1973). Weatherbee and his colleagues (1974*b*, 1975*a*) have reported, however, that human RBCs preserved in plasma and 14 per cent HES 150/0.70 during freezing yielded a saline stability value of approximately 75 per cent after thawing. Higher values of saline stability after thawing were obtained when RBCs were frozen under the same conditions but without added plasma.

Practical full-unit (375–436 ml) freezing

Allen *et al.* (1976) and Weatherbee *et al.* (1979) have reported that washing packed human RBCs with 0.9 per cent isotonic saline to remove contaminating plasma does not improve *in vitro* viability of cells previously frozen in the presence of 14 per cent HES 150/0.70. This was confirmed in earlier studies

(Weatherbee *et al.* 1975*a*) using small-unit freezing, in which the presence of plasma resulted in poor saline stability of cells when tested after thawing. The cryoprotectant medium used by most investigators contains a final concentration of 14 per cent HES 150/0.70 and varying amounts of plasma (see Appendix 6). The viability of RBCs frozen and thawed in full-units will be discussed below.

5.5.1.3. THE PROCEDURE USED TO FREEZE FULL-UNITS OF RBCs

Several investigators (Allen *et al.* 1976; Lionetti *et al.* 1975*a*, 1976; Mishler and Parry 1979; Weatherbee *et al.* 1972, 1979) have proposed various techniques for the freezing of full-units of human RBCs in the presence of HES 150/0.70. These techniques are summarized in Appendix 6.

5.5.1.4. ULTRASTRUCTURE OF RBCs FROZEN WITH HYDROXYETHYL STARCH

Small-unit (25–30 ml) freezing

Lionetti and Hunt (1974, 1975) examined the morphology of RBCs suspended in HES 150/0.70 and plasma before freezing and immediately after thawing. The *prefrozen* samples gave typical morphology of normal RBCs. However, the *thawed* samples showed a moderate number of crenated discs and echinocytic forms. After RBCs had been frozen in HES 150/0.70 without plasma, Allen and his colleagues (1977, 1978) noted that two distinct layers of cells appeared on centrifugation of thawed RBCs which had been suspended in various solutions for 4 hours or longer. Electron microscopy revealed that the *top* layer consisted of ghost-like cells nearly devoid of contents, while the *lower* layer consisted of intact, although distorted RBCs. Haematocrits of such suspended cells indicated that the top layer (ghosts) made up 12–14 per cent of the cell population. Ghost cells were not observed, however, after resuspension of thawed cells in 14 per cent HES 150/0.70 or in RBCs not suspended in various solutions but remaining in the thawed admixture of RBCs and HES 150/0.70. Electron microscopy revealed that RBCs suspended with HES 150/0.70 had major ultrastructural changes from those noted at thaw. Deposits of amorphous material appeared on the surface of a portion of the cells. Gaps or missing segments in the cell membrane were often observed with the deposits.

Full-unit (375–436 ml) freezing

Lionetti *et al.* (1976) examined the morphology of RBCs suspended in the HES 150/0.70-plasma medium before freezing, immediately after thawing, and after dilution of the unit 1:1 with glucose. The predominant cell configuration in RBCs mixed with HES 150/0.70 and plasma (precooled at 4 °C for 45 minutes) before freezing was a flat elliptical disc devoid of dimple or central pallor. Nearly all RBCs examined immediately after thawing, were flattened discs with 5–10 per cent spiculated echinocytes. These findings had been observed earlier in an additional study (Lionetti and Hunt 1975). Dehydration of the cells during freezing and rehydration during thawing contributed to the predominant configuation observed in the thawed RBCs. Reconstitution with 3 per cent glucose produced biconcave cells, stomatocytes, and echinocytes, few of which retained the flat disc shape. In marked contrast, glycerolized RBCs showed much smaller cells immediately after thawing.

Allen *et al.* (1976) examined the morphology of RBCs suspended in the HES 150/0.70 medium before freezing and immediately after thawing. Before freezing, RBC membranes were intact and continuous along the margin of the cell. Rarely, a small portion of the membrane showed a discontinuity which may have represented localized damage to the membrane. When the RBCs were fixed immediately after thawing, the membrane of most cells appeared similar to those of cells before freezing. In a small number of cells, however, the cell membrane was not continuous, but instead showed numerous gaps or regions in which a portion of the membrane was missing. These changes in the membrane after thawing were believed to represent damage acquired during the freeze–thaw process.

Allen and Weatherbee (1979) have described the appearance of RBCs frozen in the presence of varying concentrations of HES 150/0.70 using a freeze-fracture technique. The frozen RBCs appeared distorted probably as a result of osmotic dehydration but there was no evidence of intracellular ice. The frozen mixture with HES 150/0.70 had three phases; (i) a particulate phase consisting of the concentrated HES 150/0.70 (and other salts); (ii) a sculptured ice phase; and (iii) the RBCs. When the concentration of HES 150/0.70 was increased, the particulate phase came to predominate and at a concentration of 14 per cent HES 150/0.70, appeared to surround nearly all cells. In RBCs frozen in saline alone and 4 per cent HES 150/0.70, the cytoplasm in most cells had numerous cavities and depressions. Since such units haemolysed badly when thawed, it was possible that these regions indicated structural damage. In contrast, units frozen with 14 per cent HES 150/0.70 (nearly 85 per cent of cells survived the freeze–thaw process) possessed RBCs which only infrequently had such regions in the cytoplasm.

5.5.1.5. *IN VITRO* VIABILITY OF THAWED RBCs PREVIOUSLY FROZEN IN HYDROXYETHYL STARCH

Allen *et al.* (1977, 1978) had noted the ghost-like appearance of some cells when suspended after thawing in various solutions not containing HES 150/0.70 (see p. 140). Suspension of RBCs in HES 150/0.70 after thawing, however, resulted in the absence of ghosts, even though a portion of the RBC membrane possessed regions in which segments were missing. It is believed that these RBCs are among the population of cells that would eventually become ghosts in other suspension solutions. The presence of HES 150/0.70 in the external post-thaw milieu apparently serves as a barrier to prevent haemoglobin from readily diffusing into the supernatant (Scheiwe *et al.* 1979). This finding explains the discrepancy between RBC recovery and tests of saline stability. In the latter, saline affords no protection (barrier) in preventing haemoglobin from leaving damaged cells and contributing to the total external concentration. Thus many investigators believe that the saline stability test is better able to yield a more accurate measure of damage occurring to RBCs in the presence of hydroxyethyl starch (Allen *et al.* 1977; Allen and Weatherbee 1979; Scheiwe *et al.* 1979). Thus saline stability will be used to indicate the viability of RBCs after thawing.

Small unit (25-55 ml) freezing

Saline stability, cations, supernatant, and haemoglobin

Saline stability (Greenwood *et al.* 1977) and intracellular concentrations of ATP and potassium (Knorpp *et al.* 1971*a*) are not well maintained in RBCs frozen in the presence of a final concentration of 8 per cent low-viscosity hydroxyethyl starch. The saline stability of RBCs frozen in the presence of a higher concentration (14 per cent) of low-viscosity hydroxyethyl starch with or without plasma is better maintained; the value normally ranges between 75-87 per cent after thawing (Knorpp *et al.* 1971*a*; Lionetti and Hunt 1974, 1975; Weatherbee *et al.* 1975*a*). The supernatant after thawing contains 325-600 ml/dl haemoglobin, and RBCs have gained sodium (+17 per cent) and lost potassium (−14 to −21 per cent). Weatherbee *et al.* (1975*b*) claim that saline stability is improved concomitant with less free haemoglobin appearing in the supernatant after thawing when RBCs are washed free of plasma before freezing in 14 per cent HES 150/0.70 and are resuspended in saline rather than plasma after thawing. Washing previously frozen human (Knorpp *et al.* 1971*a*) or primate (Starkweather *et al.* 1971) 14 per cent HES 150/0.70 – preserved RBCs after thawing improves saline stability and reduces the liberation of free haemoglobin. Intracellular potassium, however, is reduced still further by this technique. Saline stability is reduced further after thawing concomitant with increased quantities of free haemoglobin, the longer previously frozen 14 per cent HES 150/0.70 – preserved RBCs are stored (Allen *et al.* 1978).

Intracellular enzymes and ATP

Intracellular concentrations of glucose-6-phosphate dehydrogenase (G6PD), 6-phosphogluconic acid dehydrogenase, phosphofructokinase, pyruvate kinase, glutathione reductase, lactate dehydrogenase, and hexokinase activity are maintained after thawing at control levels for a minimum of six weeks in human RBCs frozen in the presence of 14 per cent HES 150/0.70 (Starkweather *et al.* 1968). Primate RBCs frozen under similar conditions do not lose G6PD activity after thawing (Starkweather *et al.* 1971).

Intracellular concentrations of ATP after thawing are reduced 6-23 per cent in human (Baar 1973; Knorpp *et al.* 1971*a*; Lionetti and Hunt 1974, 1975) and 15 per cent in primate (Starkweather *et al.* 1971) RBCs frozen in the presence of 12-14 per cent HES 150/0.70. Intracellular concentrations of ATP appear to be better maintained in human RBCs stored in the presence of 12 per cent HES 150/0.70 plus albumin rather than 12 per cent HES 150/0.70 plus plasma

(Baar 1973). Washing human (Knorpp *et al.* 1971*a*) or primate (Starkweather *et al.* 1971) 14 per cent HES 150/0.70-preserved RBCs after thawing further reduces the ATP concentration. After 3 hours of incubation, levels of ATP in prefreeze, thawed, or thawed-washed 14 per cent HES 150/0.70-preserved RBCs was reduced 3.1, 5.4, and 7.1 per cent, respectively (Knorpp *et al.* 1971*a*).

In human RBCs frozen in the presence of 12–14 per cent low-viscosity hydroxyethyl starch, the levels of 2,3-DPG were slightly (2–10 per cent) reduced after thawing (Baar 1973; Daszyński *et al.* 1976; Lionetti and Hunt 1974, 1975). The intracellular concentration of 2,3-DPG was reduced by 20 per cent after thawing, however, in primate RBCs frozen in the presence of 14 per cent HES 150/0.70 and washed after thawing (Starkweather *et al.* 1971).

Membrane deformability

Baar (1973) measured RBC membrane deformability using a cell-filtration system. Cells processed (frozen and thawed) in 12 per cent HES 150/0.70 containing albumin had a shorter filtration time than cells frozen in 12 per cent HES 150/0.70 containing plasma.

Full-unit (375–436 ml) freezing

With the method for full-unit freezing described in Appendix 6 human RBCs frozen in the presence of 14 per cent HES 150/0.70 have post-thaw saline stabilities between 76 and 88 per cent (Table 5.12). Allen and Weatherbee (1979) reported that the final concentration of HES 150/0.70 in the freezing medium had a significant effect on RBC saline stability after thawing. Cells frozen in 14, 10, and 4 per cent HES 150/0.70 had saline stabilities of 84, 75, and 32 per cent, respectively, after thawing. Even though saline stability was relatively high in RBCs frozen in 14 per cent HES 150/0.70, these cells were distorted and did not have the normal disc shape.

Large quantities of free haemoglobin are normally found in the supernatant after thawing, which may necessitate the need for washing RBCs before transfusion (see Appendix 6, step 7).

In most experiments, slight reduction in the intracellular concentrations of both ATP and 2,3-DPG were noted in RBCs after thawing.

The oxygen delivery of RBCs after thawing is normal, and this has been substantiated in other studies (Daszyński *et al.* 1976).

The most pronounced effect of freezing RBCs in 14 per cent HES 150/0.70 is a loss of cellular potassium compensated by a corresponding increase in sodium.

5.5.1.6. *IN VIVO* SURVIVAL OF HYDROXYETHYL STARCH-CRYOPRESERVED RBCs

The survival ability of dog (Cousineau *et al.* 1972), primate (Knorpp *et al.* 1971*b*), or human (Knorpp *et al.* 1969; Weatherbee *et al.* 1979) [^{51}Cr]-labelled RBCs previously cryopreserved in 14–15 per cent low-viscosity hydroxyethyl starch to survive *in vivo* may depend on their saline stability after thawing.

TABLE 5.12 *Assessment of human RBC viability before and after a complete full-unit (375–436 ml) freeze-thaw cycle. RBCs were frozen in the presence of a final concentration of 14 per cent HES 150/0.70*

Saline stability (%)		Supernatant Haemoglobin (mg/dl)	Na^+ (mEq/1)	K^+ (mEq/1)	ATP (μmol/gHb)	2,3-DPG (μmol/gHb)	p50	Reference
Before	–	–	138	0.8	4.9	9.3	–	Lionetti and Hunt (1974, 1975)
After	76	–	105	32.4	4.7	8.3	–	
Before	99	–	165	9.2	3.2	11.2	–	Allen *et al.* (1976)
After	83	568	77	37.5	2.1	8.8	–	Weatherbee *et al.* (1979)
Before	–	–	–	–	–	–	–	Lionetti *et al.* (1976)
After	88	250–800	Lowered*	Increased†	NS	NS	Normal	Lionetti and Callahan (1979)

* By approximately 30 mEq/1
† By approximately 30 mEq/1
NS = Not significant

In dogs, the freeze–thaw process resulted in two populations of RBCs; one with normal and one with shortened times of survival. In primates, however, the lesion in RBCs induced by freezing and thawing is apparently corrected *in vivo*, resulting in the ability of cells to survive after transfusion. In humans, the survival of RBCs frozen in small aliquots (50 ml) averaged 74 per cent (range 65–89 per cent) after 24 hours (Weatherbee *et al.* 1979). These cells had an average saline stability of 85 per cent. Human RBCs frozen in full-units (560 ml) also averaged 74 per cent (range 50–92 per cent) of the infused cells after 24 hours. These cells had an average saline stability after thawing of 87 per cent. In seven patients, the 24-hour *in vivo* survival of low-viscosity hydroxyethyl starch-preserved RBCs exceeded 85 per cent of the infused cells, and the cells in an additional six patients had survivals of 75 per cent and lower (Knorpp *et al.* 1969). In two other patients, survivals of 75 and 80 per cent were observed.

5.5.1.7. THE MECHANISM OF HYDROXYETHYL STARCH CRYOPRESERVATION

Based on the studies of several investigators (Körber and Scheiwe 1977, 1980; Scheiwe 1972; Scheiwe and Krause 1977; Scheiwe and Nick 1977; Scheiwe *et al.* 1979), the presence of hydroxyethyl starch in the external freezing milieu induces certain defined physical changes (Table 5.13), which in turn are caused by a number of specific variables. The assumption that hydroxyethyl starch acts as an inert substance with respect to phase transition temperature, interfering to some extent in the transport of water by accumulation at the surface of the RBC, and therefore minimizing the effect of NaCl concentration, appears to substantiate the conclusions presented in Table 5.13. The fact that hydroxyethyl starch does not penetrate the RBC and, therefore, does not alter crystallization or minimal cell volume during freezing, contrasts with the action of either Me_2SO or glycerol. Körber and Scheiwe (1977, 1980) have indicated that unlike glycerol or Me_2SO, hydroxyethyl starch does not induce the formation of a ternary eutectic mixture when added to aqueous solutions of NaCl. Therefore, it was concluded that hydroxyethyl starch reduces the final concentration of NaCl in the residual liquid to a lower extent than either Me_2SO or glycerol.

TABLE 5.13 *Summary of the physical changes induced by the presence of hydroxyethyl starch in the freezing solution and the causes for these cryoprotective effects*

Induced physical changes	Causes
Decrease in optimal rates of cooling	Increased supercooling Smaller NaCl enrichment
Reduction of solution effects	Low diffusibility caused by viscosity Compartmentalization during freezing
Improved osmotic resistance	Fixation of the cell membrane Accumulation at the surface of the cell Aggregation of cells

*From Scheiwe *et al.* (1979)

The protective effect of hydroxyethyl starch must be viewed in the light of a combined action of influences on the physical and chemical properties of the external freezing solution (Meryman 1972). For example, the protective effect may include compartmentalization of the solution during freezing, creating a gradient of NaCl concentration around single RBCs, thus imposing different conditions of osmotic stress on the surface of frozen cells. Even though one can appreciate the numerous factors included in the freezing of suspensions of RBCs, the results of Scheiwe and his colleagues (1979) appear to agree with the two-factor theory of freeze injury for the external effects of hydroxyethyl starch.

5.5.2. GRANULOCYTES

During the past 12 years, the use of transfusion of granulocyte concentrates in the treatment of gram-negative septicaemia in patients with neutropenia has increased considerably. Presently, concentrates of transfusable granulocytes are prepared by three commonly used techniques: filtration or gravity leucapheresis, or collection by centrifugal blood-cell separators (see p. 119). The harvested concentrate of granulocytes prepared by any of the three techniques mentioned above is normally transfused several hours after collection, even though granulocytes can be stored for at least 24 and possibly 48 hours after processing if maintained at either 4 or 22 °C. The ability to store harvested concentrates of granulocytes in liquid nitrogen using cryoprotective agents for much longer periods of time under sterile conditions, without loss of function, would be extremely useful. The remaining portion of this section will discuss the results of several studies in which baboon (Lionetti *et al.* 1978), dog (French *et al.* 1980; Lionetti *et al.* 1980), rat (Bank 1980), and human (Gore *et al.* 1974; Lionetti *et al.* 1974, 1975*b*, 1980; Zaroulis and Leiderman, 1980; Zaroulis *et al.* 1978) granulocytes have been frozen in the presence of hydroxyethyl starch and Me_2SO.

5.5.2.1. BABOON GRANULOCYTES

Lionetti *et al.* (1978) isolated granulocytes from the buffy coat of baboon blood by counterflow centrifugation. Separated granulocytes were frozen (2 ml volumes containing 1×10^7 cells) in a medium consisting of 5 per cent Me_2SO, 6 per cent HES 150/0.70, 4 per cent human serum albumin, and 6 mol/l glucose in Normosol-R®. The admixture was cooled to 4 °C for 30 minutes, then cooled at 4 °C/minute to −80 °C and stored for 1-3 weeks in liquid nitrogen at −197 °C. These authors concluded that the cooling rates of 1 °C or 10 °C/minute were less efficacious. The admixture was thawed manually for 130 seconds at 42 °C in a water-bath. The recovery of granulocytes frozen by this technique averaged 98 per cent.

All baboon granulocytes isolated from fresh blood fluoresced green when incubated with fluorescein diacetate (FDA). Less than 1 per cent of fresh cells showed nuclear uptake of ethidium bromide (EB). After freezing and thawing, an average of 79 per cent of baboon granulocytes fluoresced green with FDA and 21 per cent fluoresced red with EB. Approximately 98 per cent of fresh granulocytes exhibited a capacity for ingestion of latex and yeast. Freeze–thaw granulocytes had a 75 and 67 per cent capacity to ingest latex and yeast, respectively.

5.5.2.2. DOG GRANULOCYTES

French *et al.* (1980) isolated granulocytes from the buffy coat of dog blood either by CFC or counterflow centrifugation. Separated granulocytes were frozen (1 ml volumes containing 5×10^7 cells) in a solution consisting of 5 per cent Me_2SO, 5 per cent HES 450/0.70, 2 g per cent bovine serum albumin and 20 per cent autologous citrated plasma in a modified α minimal essential medium. The admixture was cooled in a series of differing rates with final storage in liquid nitrogen vapour for 7-30 days. The frozen admixture was thawed rapidly ($\approx 75\,^\circ$C/minutes) in a water-bath. The average recovery of granulocytes frozen by this technique was 88 per cent.

Frozen, thawed, and washed dog granulocytes showed a significant decrease in chemotactic recognition and response but not chemokinetic response, although this was depressed. Phagocytosis of latex beads and the associated burst of oxygen consumption also decreased significantly to approximately 50 per cent of the original prefreeze value. However, the killing of live *E. coli* was not depressed to the extent expected and suggested by loss of oxygen consumption and selected enzyme activity.

Lionetti *et al.* (1980) isolated buffy coats of dog granulocytes from whole blood or by IFC and then further subjected these concentrates to counterflow centrifugation. Separated granulocytes were frozen and thawed under identical conditions as those described for baboon granulocytes (see p. 147). Of those cells frozen in plastic tubes, 97 per cent were recovered after thawing and displayed good stability and phagocytic indices. Cells frozen in plastic bags demonstrated nearly equal stability after eight months in liquid-nitrogen storage.

5.5.2.3. RAT GRANULOCYTES

Bank (1980) isolated granulocytes from rats by cardiac puncture or from peritoneal exudates. Separated granulocytes were frozen (a volume of 0.2 ml containing 5×10^5 cells) in a solution consisting of 10 per cent Me_2SO and 5 per cent hydroxyethyl starch in HEPES medium. The admixture was cooled by one of two methods: cooled to $-80\,^\circ$C at rates ranging from 0.3 to $15\,^\circ$C/minute and transferred directly to liquid nitrogen for storage (one-step method); or cooled to a holding temperature (ranging from -15 to $-45\,^\circ$C) and then transferred to liquid nitrogen for storage (two-step method). The admixture from either technique was thawed rapidly in a $37\,^\circ$C water-bath. Maximum functional survival averaged 71 per cent when cells were cooled at $10\,^\circ$C/minute using the one-step method, and 67 per cent when cells were held at $-40\,^\circ$C for 20 minutes using the two-step method.

In peritoneal exudate granulocytes frozen at 10 °C/minute (one-step method) in the presence of 5 per cent hydroxyethyl starch or 5 per cent hydroxyethyl starch and 10 per cent Me_2SO the chemotactic index was 66 and 45 per cent, respectively. Little difference in bactericidal activity was found between cells frozen at 10 °C/minute (one-step method) in the presence of 10 per cent Me_2SO alone or 10 per cent Me_2SO and 5 per cent hydroxyethyl starch. The quantitative NBT and FDA assays averaged 54 and 55 per cent, respectively, for peritoneal cells frozen under optimal conditions in the presence of 10 per cent Me_2SO and 5 per cent hydroxyethyl starch.

5.5.2.4. HUMAN GRANULOCYTES

Lionetti and his colleagues (Gore *et al.* 1974; Lionetti *et al.* 1975*b*, 1980) isolated buffy coats of human granulocytes from whole blood or by IFC with or without further concentration by counterflow centrifugation. Separated granulocytes were frozen and thawed under similar conditions to those previously described for baboon granulocytes (see p. 147).

Employing the freeze-thaw technique described above, the present efficiency of preservation of human granulocytes for three to four weeks of liquid-nitrogen storage is 90–100 per cent morphological and 40 per cent functional recovery.

Zaroulis and Leiderman (1980) and Zaroulis *et al.* (1978) collected buffy coats of human granulocytes by phlebotomy or by IFC and further concentrated these cells by a sedimentation/gradient technique using dextran and sodium metrizoate-Ficoll. Separated granulocytes were frozen in a medium containing a final concentration of 6 per cent HES 450/0.70, 4 per cent human serum albumin, and 5 per cent Me_2SO. The admixture was cooled at 2-3 °C/minute with maintenance at −85 °C. The admixture was thawed rapidly at 42 °C. The recovery of granulocytes frozen by this technique was 90 per cent; at least 70–89 per cent of such cells were viable by dye exclusion.

An average of 97 per cent of freshly isolated granulocytes phagocytized latex particles; in contrast, an average of 83 per cent of the same previously frozen cells demonstrated latex particle phagocytic activity.

5.5.2.5. SUMMARY

Cryopreserved granulocytes of lower animals are far superior in quality to human cells (Lionetti *et al.* 1980). Preservation efficiency is species dependent, increasing in the order human, baboon, guinea-pig, and dog. The results obtained after cryopreservation of human granulocytes in the presence of Me_2SO and hydroxyethyl starch using slow rates of cooling are encouraging; nevertheless, more work is required to make this approach more efficient. Factors affecting the stability of cryogenically preserved human granulocytes are continuing to be investigated (Lionetti *et al.* 1980) and perhaps future studies will be able to define better media or freezing techniques to adequately store human granulocytes in the frozen state.

5.5.3. PLATELETS

During the past decade, the use of platelet transfusions in the treatment of patients with thrombocytopenia has increased considerably. At present, concentrates of platelets are prepared by two commonly used methods: platelet-rich plasma is processed from either fresh blood donations or from centrifugal cytapheresis using blood-cell separators (see p. 119). In most circumstances, platelets can be stored at either 4 or 22 °C for up to 72 hours. Storage for longer periods at these temperatures has been shown to reduce the haemostatic effectiveness of collected platelets. The ability to store collected concentrates of platelets in liquid nitrogen using cryoprotective agents for a period of at least twelve months under sterile conditions without loss of haemostatic effectiveness would be an extremely useful development. I will now discuss the results of several studies in which human or primate platelets have been frozen in the presence of hydroxyethyl starch alone (Choudhury and Gunstone 1978; Kotelba-Witkowska and Gryszkiewicz 1975; Rowe and Peterson 1971; Spencer *et al.* 1969) or in combination with either Me_2SO (Kotelba-Witkowska and Gryszkiewicz 1975) or glycerol (Rowe and Peterson 1971).

5.5.3.1. *IN VITRO* STUDIES

Hydroxyethyl starch alone

Choudhury and Gunstone (1978) added *two* volumes of HES 450/0.70 to *one* volume of a cell concentrate to give a final suspension of platelet concentrate in 4 per cent HES 450/0.70. This HES 450/0.70–platelet admixture was then frozen at 1 °C/minute in a controlled-rate liquid-nitrogen freezer and stored in the vapour phase of liquid nitrogen. After storage for varying lengths of time, the admixture was thawed at 37 °C in a water-bath. *In vitro* tests on the post-thaw product indicated that 69 per cent of the original prefreeze platelets were recovered after processing. The pH of the final post-thaw concentrate averaged 6.8. In the post-thaw concentrate, over 90 per cent of platelets had undergone a shape change to spherical forms, but the platelets were discrete with relatively few microscopic aggregates. The availability of platelet factor-3 (PF-3) was higher in platelet concentrates after thawing but less in platelets stored at 4 °C for 48 hours without HES 450/0.70. The hypotonic stress response of post-thaw HES 450/0.70-preserved platelets was 54 per cent (fresh prefreeze platelets being taken as 100 per cent). Platelets stored at 4 °C for 48 hours averaged 38 per cent.

Rowe and Peterson (1971) reported that hydroxyethyl starch alone was not effective in preventing the loss of platelet function after *rapid* cooling in liquid nitrogen. After slow (1 °C/minute) and moderate (10–20 °C/minute) cooling with hydroxyethyl starch (final concentration of 8–10 per cent) protection, only 50 per cent of added [^{14}C]-serotonin was taken up by the thawed platelets.

Kotelba-Witkowska and Gryszkiewicz (1975) reported that a final concentration of 15 per cent hydroxyethyl starch was a more effective cryoprotectant than 6 per cent when platelets were cooled slowly (1 °C/minute).

Spencer *et al.* (1969) froze human platelets in a final concentration of 6 per cent HES 150/0.70. The cells were rapidly cooled at 37 °C/minutes to −80 °C and then stored at −140 °C in liquid-nitrogen vapour. *In vitro*, the post-thawed platelets had normal PF-3 activity, while clot retraction and retention of ATP was 60–70 per cent of normal.

Me_2SO and hydroxyethyl starch

When thawed platelets that had previously been slowly (1 °C/minute) frozen in the presence of 15 per cent hydroxyethyl starch and 10 per cent Me_2SO were evaluated *in vitro* by the release of ATP and ADP after thrombin stimulation and the number of cells recovered, the results were similar to those in which platelets were frozen with Me_2SO alone (Kotelba-Witkowska and Gryszkiewicz 1975).

Glycerol and hydroxyethyl starch

When platelets were frozen at slow (1 °C/minute) or moderate (10–20 °C/minute) rates, the combination of glycerol and hydroxyethyl starch prevented complete release of PF-3 after thawing (Rowe and Peterson 1971).

5.5.3.2. *IN VIVO* STUDIES

A preliminary clinical trial in five patients receiving hydroxyethyl-starch-preserved platelets (platelets suspended in 4 per cent hydroxyethyl starch and frozen slowly), has confirmed haemostatic effectiveness *in vivo* (Choudhury and Gunstone 1978). Because hydroxyethyl starch is relatively non-toxic, platelet concentrates can be infused immediately after thawing and with minimal post-thaw manipulation, thus maintaining a relatively closed system.

In primates made thrombocytopenic by irradiation, the transfusion of platelets that had been previously frozen in 6 per cent HES 450/0.70 gave good results in three out of four animals (Spencer *et al.* 1969). 17–25 per cent of platelets labelled with [^{75}Se] *in vivo* and with [^{51}Cr] *in vitro* survived, compared with 28–35 per cent of unfrozen cells.

5.5.3.3. SUMMARY

The results suggest that platelets frozen in the presence of hydroxyethyl starch alone or in combination with either Me_2SO or glycerol are preserved better if cooled at slow (1 °C/minute) or moderate (10–20 °C/minute) rates. Higher concentrations of hydroxyethyl starch (8–15 per cent) in the freezing medium appear to better preserve platelet function after thawing, even though Choudhury and Gunstone (1978) reported that final concentrations of hydroxyethyl starch in the range 5–7 per cent were detrimental to platelet function. Combinations of either Me_2SO or glycerol with hydroxyethyl starch appears to be more effective than hydroxyethyl starch alone in preserving platelet function.

5.5.4. BONE MARROW

Progress in transplantation biology has renewed the interest in bone-marrow grafting and long-term preservation of bone marrow. The remaining portion of this section will discuss the cryogenic properties of hydroxyethyl starch during the preservation of mouse (Schaefer and Beyer 1975; Schaefer *et al.* 1978), rat (Persidsky and Ellett 1971*a*, *b*), and human (Schaefer and Beyer 1975; Schaefer *et al.* 1978) bone marrow.

5.5.4.1. MOUSE BONE MARROW

Schaefer and Beyer (1975) and Schaefer *et al.* (1978) froze mouse bone-marrow cells in a medium of 17.5–20 per cent HES 450/0.70 and 20% calf serum. The cells were cooled at a rate of 1 °C/minute to approximately −40 °C. Thereafter, the cells were rapidly cooled and stored in liquid nitrogen at −196 °C for periods ranging from 2 hours to 18 months. The frozen cells were rapidly thawed in a water-bath maintained between 40 and 50 °C. With the spleen colony technique (CFU-S test) as a viability assay for haemopoietic stem cells, a mean CFU-S recovery of 89 and 78 per cent respectively, was achieved when either a rapid or slow stepwise dilution method was used to remove the cryoprotective medium.

5.5.4.2. RAT BONE MARROW

Persidsky and Ellett (1971*a*, *b*) froze rat bone-marrow cells using concentrations of 10, 15, and 20 per cent HES 150/0.70 dissolved in Hank's solution. The cells were cooled at 1 °C/minute to −33 °C with seeding at −5 °C, and then from −35 °C rapidly to −80 °C. The frozen cells were thawed rapidly in a 40 °C water-bath. Viability was assessed by the incorporation of [^{14}C]-glycine into cells before and after various regimens of freezing and thawing. Cells frozen in the presence of 15 per cent HES 150/0.70 retained approximately 22 per cent of their original glycine incorporation capacity.

5.5.4.3. HUMAN BONE MARROW

Schaefer and Beyer (1975) and Schaefer *et al.* (1978) froze and thawed human bone-marrow cells in a similar manner as was previously described for mouse cells (see above). With the agar colony technique (CFU-C test) to assess the viability of haemopoietic stem cells, a mean CFU-C recovery ranging from 60 to 80 per cent was achieved after thawing. Human bone-marrow cells can also be successfully frozen using Me_2SO and hydroxyethyl starch (Stiff *et al.* 1980).

5.5.5. LYMPHOCYTES

Mizrahi and Moore (1970, 1971*a*) froze lymphocytes in a solution containing 20 per cent calf serum, 10 per cent Me_2SO, and 10 per cent HES 150/0.70 dissolved in RPMI 1640 medium. The cells were cooled at a rate of 1–3 °C/min until −30 °C was attained, then rapid cooling was used until a temperature of

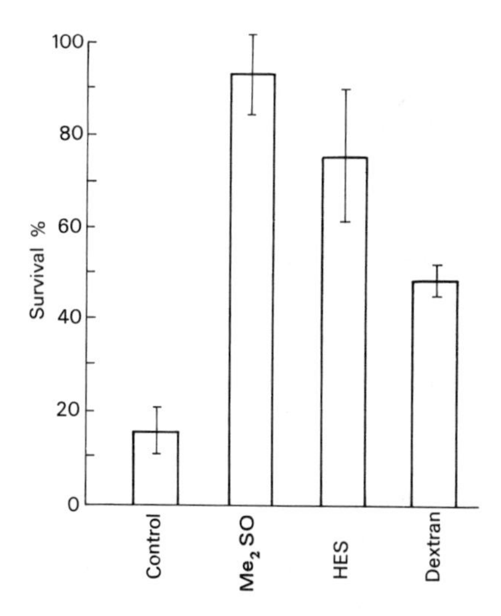

Fig. 5.3. Effect of 10 per cent Me_2SO, 10 per cent HES 450/0.70, and 10 per cent dextran 10 on survival of Chinese hamster cells frozen slowly (20 °C/minute) and rapidly thawed (115 °C/minute) in tissue-culture medium containing 10 per cent calf serum (from Ashwood-Smith *et al.* 1972*a*). Bars give standard error from mean.

−80 °C was reached. Long-term storage was at −190 °C. Cells were thawed rapidly in a 37 °C water-bath. With the freezing technique described above, over 80 per cent of lymphocytes survived. Freezing and recovery did not influence the production of immunoglobulin qualitatively or quantitatively.

These same authors (1971*b*) reported that HES 150/0.70 suspended in RPMI 1640 medium containing 2 per cent calf serum supported the rapid growth of lymphocyte cell lines. In the presence of HES 150/0.70, lower rates of glucose utilisation and lactic acid production were observed. The uptake of glucose by cells decreased as the concentration of HES 450/0.70 in the medium increased. These authors concluded that HES 450/0.70 probably protected cells against physical stress in the suspended culture.

5.5.6. CHINESE HAMSTER CELLS

Ashwood-Smith (1973) and Ashwood-Smith *et al.* (1972*a,b*) froze Chinese hamster cells in the presence of 10 per cent HES 450/0.70 or 10 per cent HES 150/0.70 in tissue culture fluid containing 10 per cent serum. When cells were frozen at 20 °C/minute and rapidly thawed (115 °C/minute), 75 per cent of cells initially frozen in the presence of HES 450/0.70 survived (Fig. 5.3). Cells frozen in the presence of HES 150/0.70 gave slightly less favourable results. Increasing the concentration of either HES 450/0.70 or HES 150/0.70 from 10 to 15 per cent did not appreciably alter results.

5.5.7. BIOLOGICAL TISSUES

Skaer and his colleagues (1978) have shown that hydroxyethyl starch can be used as a cryofixative for ultrastructural and analytical studies of biological tissues.

5.6. SUBSTITUES FOR BLOOD AND ITS COMPONENTS

Artificial blood substitutes, or more correctly, oxygen transport alternatives are of three types: synthetic chelates, stroma-free haemoglobin, and perfluorochemicals. Although synthetic chelates have not been used in biological experiments, they have the advantage of being capable of a great deal of chemical modifications. Stroma-free haemoglobin can be prepared from outdated human red blood cells, and has been used to advantage in blood replacement studies. Even though stroma-free haemoglobin has a short dwell-time in circulation, its persistence in blood can be increased by linking it chemically to hydroxyethyl starch (Geyer 1980). An entirely different approach to oxygen delivery is that achieved through the use of perfluorochemicals. Here the mechanism is simple dissolution of oxygen without the involvement of chelation, as in the case with synthetic chelators and stroma-free haemoglobin. Perfluorochemicals are composed entirely of carbon and fluorine, or also contain oxygen or nitrogen. A commercial perfluorochemical preparation, Fluosol-DA, containing perfluorodecalin and perfluorotripropylamine has been used clinically in Japan and the United States (Geyer 1978). Fluosol-DA contains hydroxyethyl starch which functions as an osmotic colloid.

5.7. PRODUCTION OF INTERFERON FROM HUMAN LEUCOCYTES

The ready availability of normal human leococytes from donated whole blood makes them an attractive source of cells for production of human interferon. Waldman and his colleagues (1981) have recently described a technique by which greater numbers of leucocytes can be processed from blood by using hydroxyethyl starch during the purification step.

Appendix 1. Estimation of hydroxyethyl starch concentration

A. HYDROXYETHYL STARCH IN AQUEOUS SOLUTION

A rapid and reliable method for estimation of hydroxyethyl starch in solution is a necessary prerequisite for any investigations with this material. Three main methods have been used: (i) polarimetric assay (Schoch 1965); (ii) evaporation to dryness followed by weighing (Greenwood and Hourston 1967); (iii) estimation of the carbohydrate by a non-specific assay such as the reaction with phenol-concentrated sulphuric acid (Dubois *et al.* 1956). All of these methods are of limited value and some cannot be applied to samples of varying molar substitution.

In the concentration range 4–40 per cent (w/v), the method of Launer and Tomimatsu (1952, 1953) appears to be rapid and reliable for estimation of hydroxyethyl starch in solution. The method described by these authors uses the complete acidification of hydroxyethyl starch to carbon dioxide and water by a mixture of potassium dichromate in hot sulphuric acid:

$$\underset{\text{starch}}{C_6H_{10}O_5} + 4\,Cr_2O_7^{2-} + 32H^+ = 6CO_2 + 21\,H_2O + 8\,Cr^{3+} \qquad \text{(A.1.1.)}$$

If an excess of dichromate is allowed to oxidize the hydroxyethyl starch and the residual dichromate is assayed, after completion of the reaction, then the amount of dichromate used is easily derived.

ASSAY PROCEDURE

A sample of hydroxyethyl starch solution (containing 0.25 ± 0.2 g dry weight polymer) is transferred to a pyrex glass beaker (400 ml) covered with a watch glass. Potassium dichromate (1.835N, 25 ml) is added. The mixture is stirred with a Teflon-coated magnetic stirrer and concentrated sulphuric acid (d.=1.84, 10 ml) is added by pipette for 10–20 seconds. After a further 15 seconds, when foaming has subsided, further acid (30 ml) is added rapidly. Water (150 ml) is added after 10 minutes and, 3 minutes later, the excess dichromate is estimated by titration with standardized ferrous ammonium sulphate (0.5 mol/1). The end-point is best determined by use of a simple electrometer (Launer and Tomimatsu 1952).

When the assay procedure was applied to dry samples of hydroxyethyl starch with molar substitution between 0.40–0.85, recoveries ranged between 99.0 and 101.0 per cent. With solutions where the concentration of hydroxyethyl starch is less than 4 per cent, the assay procedure becomes insensitive and evaporation to dryness is a more appropriate method. (Caution must be exercised that only hydroxyethyl starch is present in solution when an evaporation technique is used; most clinical solutions contain sodium chloride.)

B. HYDROXYETHYL STARCH IN BLOOD AND URINE

If hydroxyethyl starch is to be successfully estimated in blood or in urine, elaborate precautions must be taken to remove interfering substances before assay. The anthrone reaction (Appel *et al.* 1968; Roe 1954; Wallenius 1953) is most commonly used for estimation of carbohydrate in biological samples (e.g. Lindblad 1970; de Belder, personal communication), because in carefully controlled conditions, it is capable of detecting concentrations of hydroxyethyl starch as low as 30 μg/ml in serum and 90 μg/ml in urine (with a coefficient of variation of ± 5 per cent). Low-molecular-weight material, such as glucose and other sugars, interferes with the anthrone reaction. Glycogen is present only in tissue samples and is usually depleted by prior starvation of the experimental subject (de Belder, personal communication).

ASSAY PROCEDURE

Protein is removed from the sample of serum or urine by selective precipitation with trichloroacetic acid (12 per cent w/v in final solution). The hydroxyethyl starch remains in the acid supernatant and may then itself be precipitated by addition of ethanol at 4 °C. The ratio of polymer solution to ethanol should be 1:10. Extreme care is required to recover the hydroxyethyl starch precipitate, and high-speed centrifugation at low temperature is necessary for complete recovery of the precipitate. The ethanolic supernatant is decanted immediately after centrifugation and the precipitated hydroxyethyl starch resolubilised in water. Estimation of the hydroxyethyl starch in solution is then carried out by the anthrone method.

Clearly, this technique requires considerable expertise if erroneous results are to be avoided. However, a specific assay for hydroxyethyl starch in biological samples has been developed by Richter and de Belder (1976), who have prepared specific antibodies against hydroxyethyl starch. These investigators have raised antisera by immunising rabbits with a hydroxyethyl starch–bovine serum albumin conjugate. The antibodies so raised did not cross-react with starch, amylopectin, or glycogen but reacted strongly with hydroxyethyl starch (MS = 0.7–1.2). Clearly, the use of this specific antibody in modern immunological assays will provide a specific tool for estimation of low levels of hydroxyethyl starch in biological samples.

Appendix 2. Molecular weight-size distribution of hydroxyethyl starch in blood and urine

Further verification of the effect of α-amylase on the catabolism of hydroxyethyl starch may be gained by using molecular exclusion filtration (MEF). This technique allows the determination of changes taking place in the molecular weight-size distribution, either when various species of hydroxyethyl starch are incubated *in vitro* with saliva, or when these materials are allowed to circulate in the blood-stream after injection. In addition, polymer fragments of hydroxyethyl starch excreted in the urine at various intervals after dosing may be compared by MEF with material remaining in the intravascular space, thus revealing, in the general sense the threshold of the renal glomerulus. Lastly, MEF may be able to resolve the influence of either $M_{\bar{w}}$ or MS on the pattern of hydrolysis mediated by α-amylase.

TECHNICAL CONSIDERATIONS

The MEF elution pattern produced when a stock solution of HES 40/0.55 is allowed to separate on a column of either Biogel p-300 (Kono *et al.* 1972) or Sephadex G-200 (Irikura *et al.* 1972*m*) generally consists of a symmetrical, bell-shaped curve. A similar pattern of elution can be generated by passage of a stock solution of HES 450/0.70 through a column of either Sepharose CL-4B (Farrow *et al.* 1970; Mishler *et al.* 1979*a*, 1980*b*) or Sagarose 6 (Bogan *et al.* 1969). Passage of a stock solution of HES 450/0.70 through a column of either Biogel p-300 (Kono *et al.* 1972) or Sephadex G-200 (Farrow *et al.* 1970), however, normally results in a significant loss of material in the void volume (see Table A2.1 for recommended MEF columns).

TABLE A2.1 *Recommended MEF columns for the optimal separation of the various species of hydroxyethyl starch*

Species of hydroxyethyl starch	MEF column
HES 40/0.55	Biogel p-300
HES 200/0.50	Sephadex G-75 or G-200
HES 264/0.43	Sephadex G-75 or G-200
HES 350/0.60	Sepharose CL-4B
HES 450/0.70	Sagarose 6, Sepharose C-4B

HYDROLYSIS *IN VITRO* MEDIATED BY α-AMYLASE PRESENT IN SALIVA OR SERUM

When stock solutions of either HES 40/0.55 or HES 450/0.70 are incubated (at 37 °C for up to 43 hours) with saliva, the shape of the subsequently produced elution curve is altered markedly, becoming narrower and less polydispersed in comparison with the elution profile of the respective stock solution. There is a noticeable shift in the pattern of elution, indicating the presence of a greater population of polymer fragments in the small to medium molecular size range (Farrow *et al.* 1970; Irikura *et al.* 1972*m*). Incubation of the HES 450/0.70 stock solution with *serum* under similar conditions produces an intermediate change in the elution profile (Bogan *et al.* 1969).

HYDROLYSIS *IN VITRO* WITH HYDROCHLORIC ACID

An approximate bell-shaped elution profile is produced when a stock solution of high-molecular-weight hydroxyethyl starch (intrinsic viscosity: 0.286 dl/g, MS 0.54) is allowed to separate on a column of Sephadex G-75 (Tamada *et al.* 1971). Subsequent hydrolysation of this stock solution with HCl in concentrations ranging from 1/120N to 1/6N produces a family of elution patterns; the degree of polydispersion becoming an inverse function of the concentration of HCl. This inverse function is supported by parallel changes taking place in both intrinsic and relative viscosity, as well as in the amount of reducing sugar liberated.

INTRAVASCULAR HYDROLYSIS

ANIMAL MODELS

In rabbits dosed (0.6–1.8 g/kg) with either HES 40/0.55, HES, 150/0.70, or HES 450/0.70, the hydroxyethyl starch material recovered from the blood-stream 24 hours after injection was shown by MEF (Sepharose CL-4B or Sephadex G-200) to be of a narrower (less polydisperse) molecular size distribution than that of the injected stock solution (Farrow *et al.* 1970; Irikura *et al.* 1972*m*). This 24-hour post-injection elution profile indicated the presence of an *intermediate*-molecular-weight fraction occupying a position between the low- and high-molecular-weight regions of the injected stock solution (see Fig. A2.1). The influence of MS on the position of the elution profile in relation to the injected stock solution, is described in Section 2.2.2.

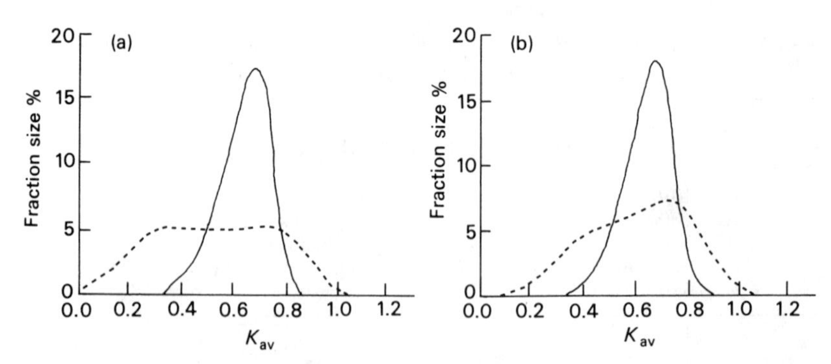

Fig. A2.1. (A). MEF on a column of Sepharose CL-4B of 50 mg of a stock solution of HES 450/0.70 (– – –) and a similar amount of hydroxyethyl starch recovered from rabbit plasma 24 hours after dosing (0.6 g/kg; ——). (B). Similar test conditions as above, except that a stock solution of HES 150/0.70 (– – –) was substituted in the infusion model and compared with the hydroxyethyl starch recovered from rabbit plasma 24 hours after dosing with this material (——). (From Farrow *et al.* 1970.)

CLINICAL MODELS

MAN

When a rapidly catabolized species of hydroxyethyl starch (MS $\leqslant$ 0.60, see Fig. 2.4) is injected into normal man, the polymers remaining in blood become progressively *smaller* (relative to the injected solution) with time (Oda *et al.* 1972*a,b*; Mishler 1980*b*; Mishler *et al.* 1979*c*, 1980*c*). However, if the MS $\geqslant$ 0.70, an *intermediate* molecular weight fraction is observed after infusion (Mishler *et al.* 1979*a*, 1980*b*). With MEF, the value of K_{av} becomes larger as a function of time after infusion in the rapidly catabolized species of hydroxyethyl starch, indicating that smaller molecules are more readily absorbed into the column, rather than being eluted into the void volume (Table A2.2). The opposite is true for the slowly catabolized species of hydroxyethyl starch (Table A2.3).

TABLE A2.2 *K_{av} values for elution peaks obtained after separation of various molecular weight material after infusion of HES 264/0.43 into normal subjects**

	Injected HES 264/0.43 solution	Postinjection period (hours)			
		0.5	1	6	24
K_{av}	0.62	0.69	0.70	0.71	0.72
K_{av}·urine/K_{av}·serum	1.34	1.20	1.19	1.17	1.15

* From Mishler *et al.* (1979*c*)

TABLE A2.3 *K_{av} values for elution peaks obtained after separation of various molecular weight material after infusion of HES 450/0.70 into normal subjects**

	Injected HES 450/0.70 solution	Postinjection period (weeks)			
		0	1	4	7
K_{av}	0.57	0.67	0.70	0.68	0.60

* From Mishler *et al.* (1980*b*)

Ultrastructural studies have shown that renal glomerular sieving is not due to the fenestrated capillary endothelium but to epithelial cells arranged so as to form interdigitating slit pores 8–10 nm wide which occupy 2–3 per cent of the total surface area (Landis and Pappenheimer 1963).

Because of the complex, branched structure of hydroxyethyl starch, it may be best to speak of the molecular radius (Stokes radius) rather than molecular weight when comparing polymers remaining in blood and those voided into urine. In this way, it is possible to document the renal threshold limit for hydroxyethyl starch. This can be documented by two examples. Firstly, a species of dextran possessing an $\bar{M_w}$ of 41 000 elutes from a column of Sepharose CL-4B with a K_{av} of 0.76, corresponding to a Stokes radius of 4.5 nm (Granath and Kvist 1967). 6–48 hours after the infusion of HES 350/0.60 in normal subjects, polymers of hydroxyethyl starch recovered from blood eluted from the same column with a K_{av} of 0.74 (Mishler 1980*c*; Mishler *et al.* 1980*c*) corresponding to an apparent molecular weight of 64 000–70 000 (Oda *et al.* 1972*a*). Although the K_{av} of dextran 40 and HES 350/0.70 were similar, the calculated molecular weights were significantly different. In this example, therefore, it can be stated that dextran 40 and HES 350/0.60 possessed a similar molecular radius but different molecular weights. Secondly, in the example cited above, it was shown that polymers of hydroxyethyl starch having a molecular radius of 4.5 nm remained in blood throughout the 6–48-hour sampling period. This would imply that the renal threshold limit for excretion of hydroxyethyl starch polymers was below a Stokes radius of 4.5 nm (9 nm diameter), which correlates with the 8–10 nm pore size as stated by Landis and Pappenheimer (1963). Thus polymers of hydroxyethyl starch possessing a molecular radius over 4.5 nm will not be filtered at the renal golmerulus.

EXCRETION *IN VIVO*

DOGS

Hydroxyethyl starch fragments contained in urine collected after dosing a dog with HES 450/0.70 were separated on a column and it was shown that the excreted polymers were of a lower molecular weight than that of the injected solution and were of a narrower range of polydispersity. There were insignificant differences in the elution profiles obtained from urine samples collected at 1, 3, and 5 hours after injection.

RABBITS

Hydroxyethyl starch recovered in the urine of rabbits dosed once with HES 40/0.55 was of a narrower molecular size distribution than that of the injected solution (Irikura *et al.* 1972*m*). The elution profiles obtained from passage of polymer fragments through a column of Sephadex G-200 were similar regardless of whether the sample was collected during the initial hour after injection, or from urine collected after 25 hours. In rabbits dosed with HES 40/0.55 on three or more consecutive days, the molecular distribution patterns of the polymers of hydroxyethyl starch contained in the collected urine indicated the presence of smaller size polymers with a slight shift to molecules of a higher molecular weight (Irikura *et al.* 1972*m*).

MAN

Polymers of hydroxyethyl starch excreted into the urine during the initial hour after injection are *smaller* than polymers recovered in urine 6–48 hours later (Oda *et al.* 1972*a*; Mishler *et al.* 1980*d*, 1981). The polymers excreted during the first hour after injection possess a K_{av} ranging from 0.88 to 0.83, which corresponded to a Stokes radius of 3.2 nm (6.4 nm diameter) (Mishler *et al.* 1980*d*, 1981). Polymers of hydroxyethyl starch excreted 6–48 hours after injection possessed a K_{av} ranging between 0.78 to 0.73, having a Stokes radius of under 4.5 nm. This observation again indicates that the renal threshold limit for excretion of hydroxyethyl starch polymers is below 4.5 nm (9 nm diameter).

During the first 8 hours after the injection of HES 40/0.55, polymers of hydroxyethyl starch voided in urine possessed an average molecular weight of 15 000–18000 (Oda *et al.* 1972*a*). However, the average *calculated* molecular weight of polymers of HES 450/0.70 excreted during this same period was approximately 58 200 (Metcalf *et al.* 1970). 12–24 hours after the infusion of either HES 40/0.55 or HES 450/0.70, the average molecular weight of excreted polymers ranged from 28 000 to 29 000 and 60 000 to 72 000, respectively.

Appendix 3. Organ perfusion

Perfusate colloids, both protein and non-protein, are used *in vitro* in the perfused kidney and heart. The beneficial effect of colloid is to maintain a physiological balance between the Starling forces at the capillaries and thereby prevent leakage of fluid from the capillaries into the interstitial spaces. With the increasing use of organ transplantation, development of suitable perfusate colloids to maintain viable organ function *in vitro* becomes an important issue (Konertz *et al.* 1976). The use of hydroxyethyl starch as a perfusate colloid in kidney and heart perfusion will now be discussed.

KIDNEY

RAT

Isolated, artificially perfused rat kidneys were perfused for 1 hour with solutions of gelatin, dextran 40, Pluronic-F-108, or HES 450/0.70 in a single-pass system (Franke *et al.* 1975). The glomerular filtration rate (GFR) during the initial 30 minutes of perfusion with either gelatin or dextran 40 was 0.58 and 0.47 ml/g/min, respectively. Using Pluronic-F-108 or HES 450/0.70, however, the GFR rose to 0.94 and 0.85 ml/g/min, respectively. With solutions of either gelatin or dextran 40, the mean tubular Na-reabsorption was 75.4 and 59.0 μmol/g/min, respectively. Using Pluronic-F-108 or HES 450/0.70, a mean net sodium transport of 92.6 μmol/g/min was achieved with either solution. The differences described in the functional capabilities of any of the four solutions during perfusion were in good agreement with morphological changes that took place in kidney ultrastructure. The most striking morphological changes were found in the proximal tubules of kidneys perfused with gelatin. On the other hand, very few morphological alterations were detected in those kidneys perfused with HES 450/0.70.

RABBIT

Isolated kidney preparations were perfused at 37 °C and an arterial pressure of 110 mm Hg using bloodless perfusates containing bovine serum albumin (BSA), dextran 70, Pluronic-F-108, or HES 150/0.70 (Fuller *et al.* 1977). The results indicated that perfusates containing BSA, dextran 70, and BSA/dextran 70 mixtures at physiological pressures did not allow normal GFR in isolated kidneys as determined by clearance of inulin (C_{in}). However, when the perfusate contained Pluronic-F-108, C_{in} was not significantly different from that found *in vivo*, and with HES 150/0.70 the measured GFR was somewhat higher. The ratio of clearance of albumin to C_{in} showed that the Pluronic-F-108 and HES 150/0.70-perfused kidneys had the lowest albumin leakage and were stable in this respect throughout a 3-hour period of perfusion. Electron microscopic examinations of glomeruli of perfused kidneys showed that the basement membrane of the glomerular capillaries was intact with all perfusates and no evidence of morphological changes were noted, but structural changes were evident in the capillary endothelial lining. In kidneys perfused with BSA or dextran 70/BSA there was a distinctive swelling and coalescence of the endothelium, with a consequent partial occlusion of the endothelial fenestrae in most capillary loops examined. Extensive sections of endothelial lining were missing in kidneys perfused with dextran 70 alone and HES 150/0.70. Moderate fusion of the epithelial-cell foot processes were common in kidneys perfused with all solutions, and large balloon-like vacuoles in the visceral epithelial cells – which encroached extensively into the urinary space between the capillary loops – were a distinctive structural alteration in kidneys perfused with either dextran 70 alone or with HES 150/0.70. Even with these structural alterations, HES 150/0.70 gave high GFR and dextran 70 yielded low GFR. The discrepancy between altered ultrastructure and normal organ function *in vivo* has been discussed previously (Section 2.4 and Section 3.9).

When kidneys treated with cyanide and iodoacetate were perfused at 37 °C with an initial arterial pressure of 60 mm Hg that was later reduced to 40 mm Hg, total water content increased significantly in kidneys perfused with HES 150/0.70 (Pegg 1977). The EDTA space increased during perfusion with HES 150/0.70, as did the albumin space. Kidneys gained sodium and lost potassium. Normal vascular resistance, however, was maintained. In contrast to the results of Pegg (1977), Jeske and his colleagues (1974*a*, *b*), adopting a different mode of perfusion, found that weight gain (oedema) was greatest in kidneys perfused without colloid, and lower, albeit insignificantly, in kidneys perfused with HES 150/0.70. In these experiments, vascular resistance was substantially reduced by perfusion with HES 150/0.70, and this was reflected in significantly greater perfusate flow rates in comparison with perfusion without colloid. Despite reduced vascular resistance values in the HES 150/0.70-perfused organs, clearance of creatinine was considerably lower than that observed in kidneys perfused without colloid. Urine flow rates were significantly greater in kidneys perfused without colloid than in those perfused with either dextran 70 or HES 150/0.70. The decrease in urine flow rates was quite pronounced in the dextran 70-perfused kidneys. Fractional sodium reabsorption was significantly greater in kidneys perfused with HES 150/0.70 than in perfusates containing no colloid at 30 minutes, but the differences were not significant at 1 hour.

In studies in which isolated kidneys were perfused at 37 °C with an arterial pressure of 110 mm Hg, HES 450/0.70-perfused kidneys gave similar GFR when compared with Pluronic-F-108 and gelatin, but higher values were achieved with dextran 150 (Wusteman 1978). Protein leakage was highest with HES 450/0.70 and least with dextran 150 and gelatin.

HEART

RABBIT

Isolated rabbit hearts were perfused with various electrolyte or colloid solutions under normothermic conditions. In these circumstances, both heart rate and contraction amplitude were immediately reduced on introducing HES 450/0.70 (suspended in Krebs–Henseleit bicarbonate solution) to the heart (Armitage and Pegg 1977). A 2:1 heart block quickly developed and the ECG became flat very soon after the chambers had ceased beating. The tissue appeared very oedematous.

DOG

Cadaveric hearts or hearts obtained from living dogs were either preserved or perfused with a solution of HES 450/0.70 with normal or greater than normal amounts of oxygen (Garzon *et al.* 1967*b*). Cadaveric hearts preserved at 4 °C for 24 hours in HES 450/0.70 with normal amounts of oxygen fibrillated but did not regain rhythmic contractions when perfused with blood on rewarming. The activities of alanine aminotransferase (ALT) increased from 193 to 616 IU/g heart tissue. Cadaveric hearts perfused with HES 450/0.70 and normal quantities of oxygen regained rhythmic contractions; however, the mean maximal rate of left ventricular pressure rise (dp/dt) decreased from 578 to 160 mm Hg/sec, and ALT increased from 186 to 245 IU/g heart tissue. Cadaveric hearts preserved in HES 450/0.70 with greater than normal quantities of oxygen could be defibrillated; the mean dp/dt decreased from 520 to 162 mm Hg/s, and ALT rose up to 269 IU/g heart tissue. Hearts obtained from living dogs and preserved in HES 450/0.70 with normal quantities of oxygen regained rhythmic contractions on rewarming, with a mean dp/dt of 280 mm Hg/s concomitant with an increase of ALT from 113 to 199 IU/g heart tissue. The authors concluded that cadaveric hearts would regain rhythmic contractions only when perfused or preserved with increased quantities of oxygen. ALT levels in the preserving solution correlated with dp/dt and may be useful for predicting viability.

Appendix 4. The technique of gravity leucapheresis

(1) Venous catheters are inserted into large veins on both donor arms. Blood is withdrawn from the first catheter while the catheter located in the remaining arm is used to re-infuse RBCs and plasma after granulocyte processing (Fig. A4.1).

(2) 500 ml of blood are collected into a standard PVC–plastic bag (A) containing 75 ml ACD-A.

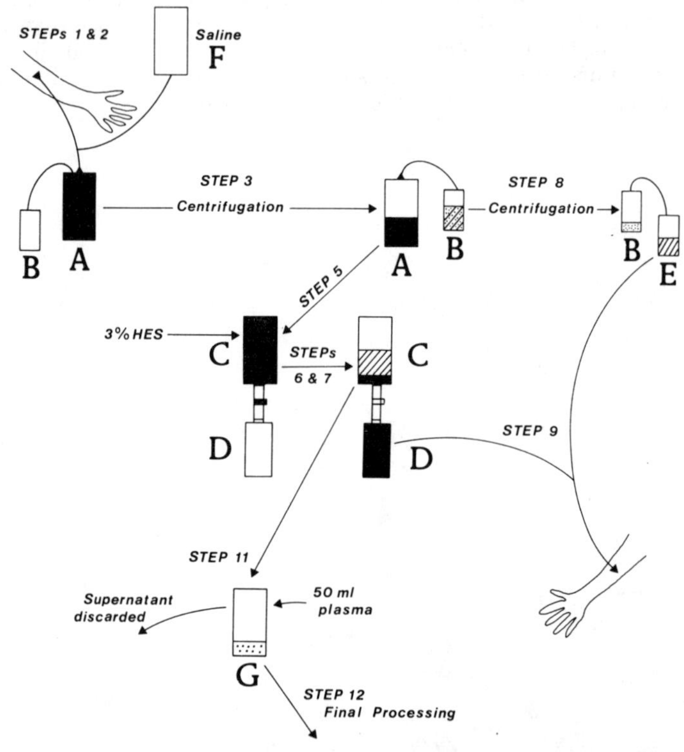

Fig. A4.1. A diagrammatic representation of the steps in preparing both platelets (bag B) and granulocytes (bag G) by batch-processing of donor units sedimented with 3 per cent (w/v) HES 450/0.70–0.9 per cent isotonic saline solution. The 0.9 per cent isotonic saline contained in bag F is used to maintain free flow in the collection vein (steps 1 and 2). The procedure can be accelerated in large donors able to tolerate the consecutive collection of two donor units in two separate ACD-A bags (2 × step 2). In such circumstances the two donor units are processed separately but simultaneously as described above. Bleeding of the donor is resumed after the return of at least one unit (plasma and RBCs – step 9) to the donor. (From Djerassi 1977.)

(3) The collected donor unit is centrifuged for 6 minutes (at 2000 rpm) at room temperature.
(4) The platelet-rich plasma (PRP) is expressed (with the aid of a spring-back plasma extractor) into a PVC-satellite bag (B).
(5) The buffy coat and packed RBCs are aseptically transferred to a 2-litre PVC-sedimentation bag (C). HES 450/0.70 (3 per cent w/v) in 0.9 per cent isotonic saline is added to the buffy coat–RBC admixture to achieve an approximate haematocrit of 30 per cent (*Note*: usually 500 ml of 3 per cent HES 450/0.70–saline solution can be added without determining the haematocrit, providing adequate results.) The resulting admixture is mixed several times by repeated inversion and then is suspended (at 90°) with the outlet and inlet tubings at the bottom.
(6) A 300 ml PVC-transfer bag (D) is attached to the 2-litre sedimentation bag (C) and the tubing clamped until the RBCs are sedimented in the sedimentation bag (C).
(7) RBCs are then drained by gravity into a PVC-transfer bag (D) leaving behind the uppermost 10–20 ml of RBCs in the sedimentation bag (C) together with the supernatant containing HES 450/0.70–saline and leucocytes. (*Note*: separation of the RBCs by gravity alone is usually completed in 15–20 minutes.)
(8) During this RBC-sedimentation period (steps 5–7), the PRP initially separated and collected into the PVC-satellite bag (B) (step 4) is centrifuged (at 2800 rpm for 10 min). The platelet-poor plasma (PPP) is collected into a PVC-transfer bag (E).
(9) The PPP (Bag E, step 8) and RBCs (Bag D, step 7) are re-infused into the donor via the *second* indwelling catheter.
(10) While returning the plasma and RBCs (step 9), a second unit of donor blood is collected (repeated of step 2).
(11) The sedimentation bag (C) containing the HES 450/0.70–saline–leucocyte supernatant (step 7) is centrifuged immediately at 600 rpm for 18 minutes. The supernatant HES 450/0.70 solution is discarded, leaving behind some contaminating RBCs and the granulocytes in the sedimentation bag (C). Approximately 50 ml normal ABO-compatible plasma is introduced into the sedimentation bag (C) and the granulocytes resuspended by repeated inversion.
(12) After the processing of the desired number of donor units (repeating step 2), the granulocytes contained in the numerous processed sedimentation bags (C) (step 11) are pooled into a single 300 ml PVC-transfer bag. The RBCs are further sedimented by the addition of HES 450/0.70 (6 per cent (w/v) in 0.9 per cent isotonic saline) in a ratio sufficient to produce a final concentration of 1 per cent. The 300 ml PVC-transfer bag is then hung upside down (at 90°) with the outlet and inlet tubings at the bottom. After complete sedimentation, the RBCs (up to 100 ml) are drained into a separate transfer bag and returned to the donor via the *second* indwelling catheter. The final concentrate contains granulocytes and some contaminating platelets.

Appendix 5. The preparation of leucocyte-poor blood (LP–RBC)

(1) 450 ml of donor blood are obtained by venepuncture and collected into a standard PVC-plastic bag containing 63 ml CPD.
(2) Sedimentation is performed in a 1-litre PVC-bag (sedimentation bag) with three spike lines and an attached PVC-satellite bag. *One* spike line is connected to a 500 ml container of 6 per cent (w/v) HES 450/0.70 in 0.9 per cent isotonic saline. A *second* spike line is attached to a 1-litre bag of 0.9 per cent isotonic saline. The *third* spike is connected to the unit of previously collected whole blood (step 1). The unit of whole blood is then allowed to drain into the empty 1-litre sedimentation bag.
(3) Approximately 30 ml 0.9 per cent isotonic saline are then introduced into the original (now empty) unit of blood, and this rinse is then combined with the blood contained in the 1-litre sedimentation bag (step 2).
(4) The spike line to the original unit of blood (now rinsed free of RBCs) is heat-sealed close to the donor unit, clamped, and cut just proximal to the seal. This spike line is subsequently used as the exit line for discarding supernatant fluid after sedimentations.
(5) Equal volumes of HES 450/0.70 and saline are allowed to fill the 1-litre sedimentation bag containing blood (step 2), and the subsequent admixture is then thoroughly mixed by repeated inversion.
(6) The 1-litre sedimentation bag containing the blood–HES 450/0.70 admixture is then positioned at an angle of 60° with the outlet and inlet tubings at the bottom. After 25 minutes, the 1-litre sedimentation bag is hung upside down (at an angle of 90°) for a further 5 minutes (10 minutes if sedimentation appears incomplete) and the sedimented RBCs are allowed to drain out by gravity into the attached satellite (final transfusion container), leaving behind the uppermost layer of RBCs (approximately 10–15 ml) to be discarded with the HES 450/0.70 saline solution.
(7) The RBCs (contained in the attached satellite bag) are reintroduced into the 1-litre sedimentation bag (now rinsed free of contaminating RBCs) and step 6 is repeated a second time, again returning all but the uppermost RBCs to the final transfusion container. Dorner *et al.* (1975) have proposed an integral unit (see Fig. A5.1) for the preparation of HES 450/0.70 LP-RBCs, which would reduce both wastage of solutions and plastics whilst minimizing the time required for preparation of the final product.

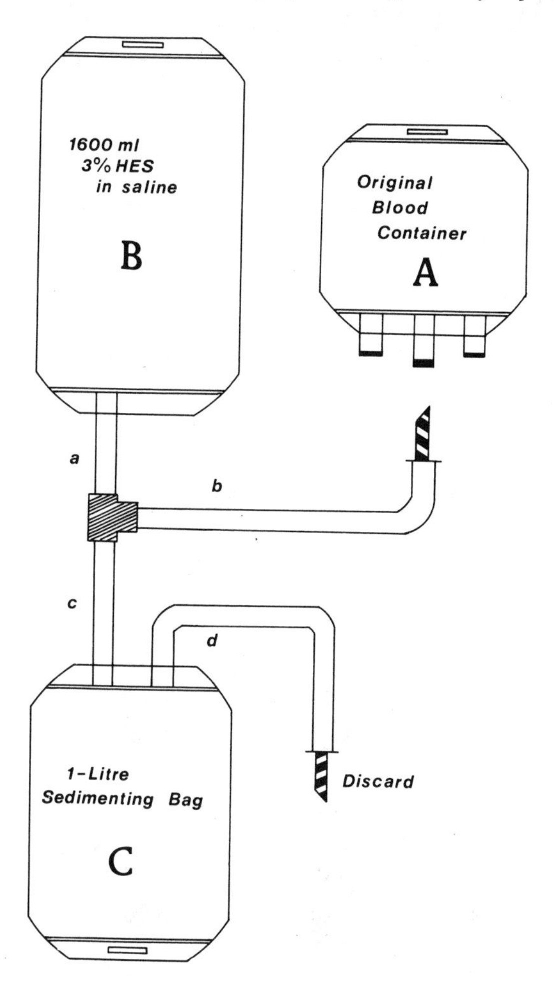

Fig. A5.1. Diagram of a proposed integral unit for preparation of LP-RBCs by HES 450/0.70 sedimentation (Dorner *et al.* 1975). With the proposed technique the LP-RBCs would ultimately be returned to the original blood container (A), which would remain attached until completion of the procedure. Positions *a*, *b*, *c*, and *d* indicate points at which tubings would be clamped at various stages of processing, i.e. points *a* and *d* clamped during admission of original blood to sedimentation bag (step 2); points *c* and *d* clamped during the addition of HES 450/0.70–saline solution (B) to the original container (A) to *rinse* in residual donor blood (step 3); points *b* and *d* clamped during the addition of the HES 450/0.70–saline solution (B) to the sedimentation bag (C) (step 5); and *c* and *b* clamped during discard of supernatant after sedimentation (step 6).

Appendix 6. Full-unit freezing of RBCs with hydroxyethyl starch

(1) 450 ml of human whole blood are collected into a PVC plastic blood bag containing 63 ml CPD (Fig. A6.1).

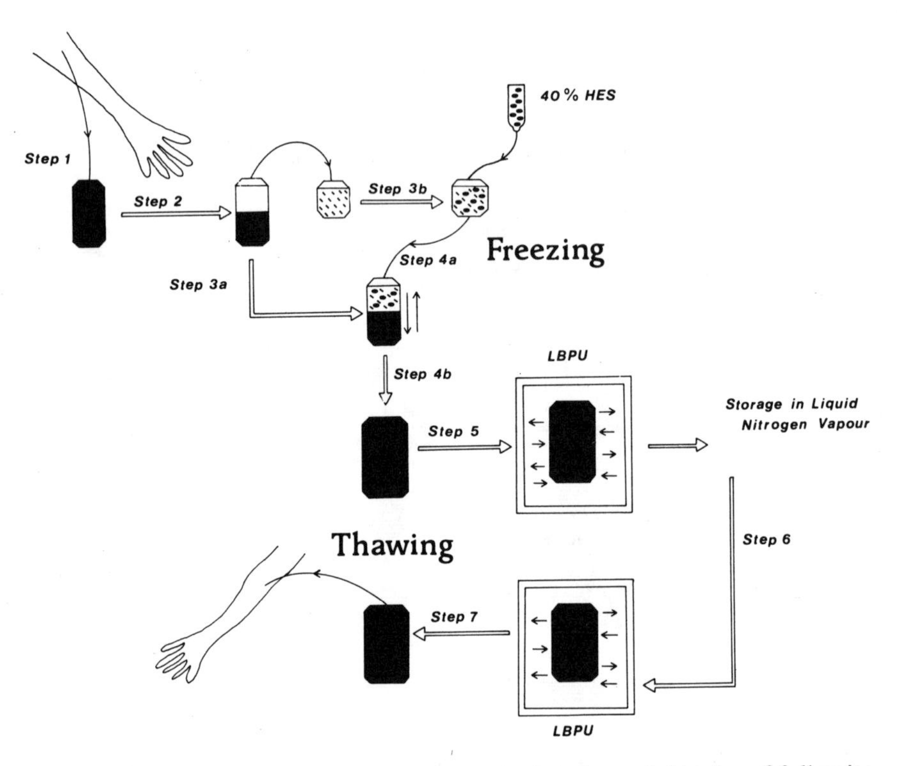

Fig. A6.1. Schematic summary of the steps in freezing and thawing of full-units (375–436 ml) of RBCs with HES 150/0.70.

(2) From the citrated whole blood, suspensions of packed RBCs are prepared by centrifugation (1800–2750 *g*, 5–15 minutes at 0–4 °C) followed by expulsion of the supernatant plasma into an ancillary PVC plastic blood bag.

(3a) The remaining packed RBCs are adjusted to: (i) a weight of 200 g (Mishler and Parry 1979) and either (ii) a haematocrit of approximately 40 per cent by the addition of plasma or 0.9 per cent isotonic saline (Allen *et al.* 1976; Weatherbee *et al.* 1979); or (iii) a constant packed RBC–plasma volume of 263 ml (Lionetti *et al.* 1976; Lionetti and Callahan 1979).

(3b) 140 ml 40 per cent (w/v) HES 150/0.70 are added slowly to 60 ml of autologous plasma (Mishler and Parry 1979).

(4a) The HES 150/0.70–plasma mixture is carefully agitated and then added slowly to the packed RBCs.

(4b) The full-unit of RBCs is carefully inverted several times and then transferred aseptically to a freezing bag (Union Carbide, USA). The full-unit is then kept at 4 °C for 45–120 minutes before freezing (Allen *et al.* 1976; Lionetti *et al.* 1976).

(5) The full-unit is then placed in a Linde Blood Processing Unit (LPBU) and automatically immersed and shaken at either 200 cycles/minute for 90 seconds (Mishler and Parry 1979), or 200 cycles/minute for 60 seconds followed by 150 cycles/minute for an additional 60 seconds. The latter two-stage freezing cycle has been shown to reduce bag breakage on thawing (Allen *et al.* 1976). The frozen full-unit is then stored in liquid-nitrogen vapour at −140 °C.

(6) The frozen unit is thawed by rapid immersion and agitation at: (i) 160 cycles/minute at 55 °C for 60 seconds (Mishler and Parry 1979) (ii) 200 cycles/minute at 54 °C for 60 seconds (Lionetti *et al.* 1976) or (iii) 170 cycles/minute at 47–49 °C for 60–70 seconds (Allen *et al.* 1976).

(7) The thawed unit of RBCs can be directly transfused (Mishler and Parry 1979), or transfused after an initial wash with either 6 per cent glucose (Lionetti *et al.* 1976) or 0.9 per cent isotonic saline (Allen *et al.* 1976; Knorpp *et al.* 1971*a*; Weatherbee *et al.* 1979).

References

Abe, H. (1976). Maintenance – effect of blood pressure by Hespander for prevention of hypotension at spinal anesthesia. *Shinryo-to-Shinyaka* **13**, 517.

Adachi, K., Samejima, T., Honda, Y., and Yoshitake, J. (1972). Clinical experiences with Hespander with special reference to its effect on metabolism and insulin excretion. *Rinsho-to-Kenkyu* **49**, 276.

Aisner, J., Schiffer, C.A., Wolff, J.H., and Wiernik, P.H. (1976). A standardized technique for efficient platelet and leukocyte collection using the Model 30 Blood Processor. *Transfusion* **16**, 437.

—— —— Hedmann, S., and Wiernik, P.H. (1979). Gravity leukapheresis – a simple method of leukocyte collection for transfusion or in vitro study. *Leukemia Res.* **3**, 1.

—— —— —— and Wiernik, P.H. (1977). Gravity leukapheresis – a simple method of leukocyte collection for transfusion or in-vitro study. *Blood* **50** Suppl. 1, 304.

Alexander, B. and Odake, K. (1967). Cryoprecipitation of factors 1 and 8 by dextran (D) and hydroxyethyl starch (HES). *Fedn Proc. Fedn Am. Socs exp. Biol.* **26**, 487.

—— —— Lawlor, D., and Swanger, M. (1975). Coagulation, hemostasis, and plasma expanders. A quarter century enigma. *Fedn Proc. Fedn Am. Socs exp. Biol.* **34**, 1429.

Allen, E.D. and Weatherbee, L. (1979). Ultrastructure of red cells frozen with hydroxyethyl starch. *J. Microsc.* **117**, 381.

—— —— and Permoad, P.A. (1977). Post-thaw suspension of red-cells cryopreserved with hydroxyethyl starch. *Cryobiology* **14**, 708.

—— —— —— (1978). Post-thaw suspension of red-cells cryopreserved with hydroxyethyl starch. *Cryobiology* **15**, 375.

—— —— Spencer, H.H., Lindenauer, S.M., and Permoad, P.A. (1976). Large unit red cell cryopreservation with hydroxyethyl starch. *Cryobiology* **13**, 500.

Alsop, R.M., Byrne, G.A., Done, J.N., Earl, I.E., and Gibbs, R. (1977). Quality assurance in clinical dextran manufacture by molecular weight characterization. *Process Biochem.* **12**, 15.

Amakata, Y., Tanaka, K., Hanai, A., and Kosaki, G. (1972*a*). Clinical use of a new plasma expander, hydroxyethyl starch during anesthesia. III. Statistical comparisons between hydroxyethyl starch and dextran 70 on blood pressure, blood analysis, electrolytes and hemostasis. *Jap. J. Anesthesiol.* **21**, 225.

—— —— —— —— (1972*b*). Clinical use of a new plasma expander, hydroxyethyl starch during anesthesia. IV. Statistical comparison between hydroxyethyl and dextran 70 on serum enzymic activities, hepatic functions, erythrocyte sedimentation rates and renal function. *Jap. J. Anesthesiol.* **21**, 361.

—— —— Kosaki, G., Hara, M., and Umemoto, Y. (1971). Clinical use of a new plasma expander, hydroxyethyl starch during anesthesia. I. Effects on blood pressure, blood analysis, electrolytes and hemostasis. *Jap. J. Anesthesiol.* **20**, 1251.

—— —— —— —— —— —— (1972*c*). Clinical use of the new plasma expander, hydroxyethyl starch during anesthesia. II. Examination of serum enzymic activities, hepatic functions, erythrocyte sedimentation rate, renal functions and untoward effects. *Jap. J. Anesthesiol.* **21**, 40.

Anderson, D.M.W. and Zaidi, S.S.H. (1963). Applications of infrared spectroscopy

XII. The behaviour of propoxyl and butoxyl groups in the zeisel reaction. *Talanta* **10**, 1235.

Anderson, G.H., Robbins, F.M., Domingues, F.J., Moores, R.G., and Tong, C.L. (1963). The utilization of Schardinger dextrins by the rat. *Toxicol. Appl. Pharmacol.* **5**, 257.

Appel, V.W., Werner, V., and Sprengard, D. (1968). Quantitative microestimation of dextran. *Z. Chem. Klin. Biochem.* **6**, 452.

Arita, H. and Matsushima, Y. (1970). Studies on the substrate specificity of Taka-amylase A. VI. Synthesis of phenyl 2′-O-methyl-α-maltoside and as enzymatic investigation. *J. Biochem.* **68**, 712.

—— —— Isemura, M., and Irenaka, T. (1970). Studies on the substrate specificity on Taka-amylase A. IV. The mode of action of Taka-amylase A on modified phenyl α-maltoside. *J. Biochem.* **68**, 91.

Armitage, W.J. and Pegg, D.E. (1977). An evaluation of colloid solutions for normothermic perfusion of rabbit hearts: An improved perfusate containing Haemaccel. *Cryobiology* **14**, 428.

Arrants, J.E., Cooper, N., and Lee, W.H. (1969). The effects of a new plasma expander (hydroxyethyl starch) on intravascular clot formation. *Am. Surg.* **35**, 465.

Ashwood-Smith, M.J. (1973). Problems of cell survival after freezing and thawing with special reference to the cornea. In *Corneal Graft Failure. Ciba Foundation Symposium 15 (New Series)*, pp. 57–77. Elsevier-Excerpta Medica, Amsterdam.

—— Warby, C., Becker, G., and Connor, K.W. (1972*a*). Protective action of various polymers, including polyvinylpyrrolidone, dextran and hydroxyethyl starch against freezing damage to mammalian cells in tissue culture. *Cryobiology* **9**, 311.

—— —— —— —— (1972*b*). Low-temperature preservation of mammalian cells in tissue culture with polyvinylpyrrolidone (PVP), dextrans, and hyroxyethyl starch (HES). *Cryobiology* **9**, 441.

Baar, S. (1973). Albumin and hydroxy-ethyl starch in the cryopreservation of red cells – an in vitro. *Transfusion* **13**, 73.

Ballinger, W.F. (1965). The comparative effect of hydroxyethyl starch, dextran, and whole blood on survival of dogs subjected to hemorrhagic shock. In *Proceedings of the Third Conference on Artificial Colloidal Agents*, p. 158 NAS-NRC, Washington, DC.

—— Murray, F.G. and Morse, E.E. (1966*a*). Preliminary report on the use of hydroxyethyl starch solution in man. *J. Surg. Res.* **6**, 180.

—— Solanke, T.F., and Thompson, W.L. (1966*b*). The effect of hydroxyethyl starch upon survival of dogs subjected to hemorrhagic shock. *Surg. Gynecol. Obstet.* **122**, 33.

Bank, H.L. (1980). Viability of frozen rat granulocytes and granulocyte precursors. *Cryobiology* **17**, 262.

Banks, W. and Greenwood, C.T. (1975). *Starch and its components.* Edinburgh University Press.

—— —— and Muir, D.D. (1970*a*). The characterization of starch and its components: the semi-micro estimation of the starch content of cereal grains and related materials. *Die Staerke* **22**, 105.

—— —— and Walker, J.T. (1970*b*). The starches from waxy genotypes of barley. *Die Staerke* **22**, 149.

—— —— and Muir, D.D. (1971). Report to US Office of Naval Research, No. 00014-70-C-0101.

—— —— —— (1972). Studies on hydroxyethyl starch. *Die Staerke* **24**, 181.

—— —— —— (1973). Structure of hydroxyethyl starch. *Br. J. Pharmacol.* **47**, 172.
Bearden, J.D., Ratkin, G.A., and Coltman, C.A. (1975). Hydroxyethyl starch (HES) and prednisone (P) as adjuncts to granulocyte collection. *Clin. Res.* **23**, A336.
—— —— —— (1977). Hydroxyethyl starch and prednisone as adjuncts to granulocyte collection. *Transfusion* **17**, 141.
Beez, M. and Dietl, H. (1979). Retrospektive Betrachtung der Häufigkeit anaphylaktoider Reaktionen nach Plasmasteril and Longasteril. *Infusionsther. Klin. Ernaehr.* **6**, 3.
Bendavid, A. and Gavendo, S. (1971). Protection of red cells against noncryogenic injury by polyvinylpyrrolidone and hydroxyethyl starch. *Cryobiology* **8**, 381.
—— —— (1972). Protective effect of polyvinylpyrrolidone and hydroxyethyl starch on noncryogenic injury to red blood cells. *Cryobiology* **9**, 192.
Bergmann, H., Blauhut, B., and Necek, S. (1978). Die Verwendung von Hydroxyaethylstaerke bei der Zell-separation. *Wien. Med. Wochenschr.* **128**, 333.
Berman, H.J., Niessen, N.E., and Fulton, G.P. (1965). Hemostatic responses in hamsters 24 hours after infusion with hydroxyethyl starch: A preliminary report. In *Proceedings of the Third Conference on Artificial Colloidal Agents*, p. 99. NAS-NRC, Washington, DC.
Bode, J.C., Krupinski, R., and Dürr, H.K. (1977). Comparison of naturally occurring macroamylasemia with macroamylasemia induced by intravenous-infusion of hydroxyethyl starch (HES). *Ir. med. J.* **146**, 23.
Bode, V. and Deisseroth, A.B. (1981). Donor toxicity in granulocyte collections: association of lichen planus with the use of hydroxyethyl starch leukapheresis. *Transfusion* **21**, 83.
Bogan, R.K., Gale, G.R., and Walton, R.P. (1969). Fate of ^{14}C-labelled hydroxyethyl starch in animals. *Toxicol. appl. Pharmacol.* **15**, 206.
Bollenback, G.N., Golik, R.S., and Parrish, F.W. (1969). Distribution of hydroxyethyl starch. *Cereal Chem.* **46**, 304.
Boon, J., Jesch, F., Ring, J., and Messmer, K. (1976). Intravascular persistence of hydroxyethyl starch in man. *Eur. Surg. Res.* **8**, 497.
Borberg, H. (1978). Leucocyte collection using continuous flow centrifugation. In *Cell-separation and cryobiology* (ed. H. Rainer *et al.*). Schattauer, Stuttgart.
Boxall, L.M., Greenwood, C.T., and Muir, D.D. (1974). An effective gas-absorption trap for use in the analyses of hydroxyethyl ethers. *Lab. Pract.* **23**, 67.
Braine, H.G., Elfenbein, J., and Scribner, P. (1980). Effect of repeated cytapheresis on in vitro lymphocyte reactivity in normal donors. *Transfusion* **20**, 649.
Brammer, G.L., Rouguie, M.A., and French, D. (1972). Distribution of α-amylase-resistant regions in the glycogen molecule. *Carbohydr. Res.* **24**, 343.
Brickman, R.D., Murray, G.F., Thompson, W.L., and Ballinger, W.F. (1966*a*). Investigation of antigenicity of hydroxyethyl starch in humans. *J. Am. med. Ass.* **196**, 575.
—— —— —— —— (1966*b*). The antigenicity of hydroxyethyl starch in humans. Studies in seven normal volunteers. *J. Am. med. Ass.* **198**, 1277.
Brittingham, T.E. and Chaplin, H. (1957). Febrile transfusion reactions caused by sensitivity to donor leukocytes and platelets. *J. Am. med. Ass.* **165**, 819.
Broderick, A.E. (1954). Hydroxyalkylation of polysaccharides. US Patent 2682 535.
—— (1960). Granulated, water-soluble hydroxyalkyl starch. German Patent 1019 288.

Brown, T.G., Evangelista, B.S., Green, T.J., and Gwilt, D.J. (1968). Experimental study of current therapeutic approaches to endotoxin shock. *Fedn Proc. Fedn Am. Socs exp. Biol.* **27**, 447.

Burchard, W. and Cowie, J.M.G. (1972). *Light scattering* (ed. M. Hughlin. Academic Press, New York.

Caldwell, C.G. and Martin, I. (1957). Ungelatinized cold-water soluble starch ethers. US Patent 2 802000.

Cerny, L.C., Graham, R.C., and Daniels, C.A. (1965). Physicochemical properties of hydroxyethyl starch. In *Proceedings of the Third Conference on Artificial Colloidal Agents.* p. 20. NAS-NRC, Washington, DC.

—— —— and James, H. (1967). Dilute solution properties of hydroxyethyl starch. *J. appl. Polymer Sci.* **11**, 1941.

—— Granz, J.D., and James, H. (1968). Dilute solution properties of hydroxyethyl starch. *Biorheology* **5**, 103.

Chan, Y.C. (1975). 0-(2-hydroxyethyl)-Amylose as the Substrate of Porcine Pancratic α-Amylase Action: Structural Analysis of 0-(2-hydroxyethyl)-Maltooligosaccharides. Ph.D. Thesis, Iowa State University.

Cheng, C., Lerner, B., Lichtenstein, S., Karlson, K.E., and Garzon, A.A. (1966). Effect of hydroxyethyl starch on hemostasis. *Surg. Forum* **17**, 48.

Chien, S., Dellenback, R.J., Usami, S., and Gregersen, M.I. (1965). Capillary permeability to macromolecules in endotoxin shock. In *Proceedings of the Third Conference on Artificial Colloidal Agents* p. 56. NAS-NRC, Washington, DC.

Choudhury, C. and Gunstone, M.J. (1978). Freeze preservation of platelets using hydroxyethyl starch (HES): A preliminary report. *Cryobiology* **15**, 493.

Chu, C.C. (1979). Experimental studies on hydroxyethyl starch as a diluent for artificial heart lung machine performance. *Taiwan I Hsueh Hui Tsa Chih* **78**, 784.

Ckiba, T. (1974). Experience of the use of Hespander (HES) during hypothermal anesthesia (mainly in cases of open heart surgery). *J. New Rem. Clin.* **23**, 1537.

Cousineau, L., Forest, R., and Longpre, B. (1972). In-vivo ^{51}Cr survivals of dog erythrocytes frozen in hyroxyethyl starch (HES). *Cryobiology* **9**, 317.

Daszyński, J., Gryszkiewicz, A., and Kościekak, J. (1976). Erythrocyte freezing in the presence of hydroxyethyl starch. *Acta haematol. pol.* **7**, 155.

de Belder, A.N. and Norrman, B. (1969). Substitution patterns of O-(2-hydroxyethyl) starch and O-(2-hydroxyethyl) dextrin. *Carbohydr. Res.* **10**, 391.

—— Markstrom, S., and Perrson, A. (1972). Critical features of the Morgan assay of hydroxyethyl and hydroxypropyl substitutes in polysaccharides. *Die Staerke* **24**, 361.

—— —— and Larsson, S.O. (1976). Molecular weight, substitution and impurity studies of some hydroxyethyl starch plasma volume expanders. *IRCS Med. Sci. Drug Metabolism Toxicol. Pharmacol.* **4**, 457.

Dillon, J., Lynch, L.J., Myers, R., Butcher, H.R., and Moyer, C.A. (1966). A bioassay of treatment of hemorrhagic shock. I. The roles of blood, Ringer's solution with lactate, and macromolecules (dextran and hydroxyethyl starch) in the treatment of hemorrhagic shock in the anesthetized dog. *Archs Surg.* **93**, 537.

DiMarco, J.P., Bloxham, D.D., and Thompson, W.L. (1978). Low-molecular weight hydroxyethyl starch – kinetics in man. *Clin. Res.* **26**, A288.

Djerassi, I. (1977). Gravity leucapheresis – A new method for collection of transfusable granulocytes. *Exp. Hematol.* **5** (Suppl. 1), 139.

—— Goldman, J.M., and Murray, K.H. (1977). Standards for filtration leucapheresis – A prerequisite for further development. *Exp. Hematol.* **5** Suppl. 1, 49.

Doenicke, A., Grote, B., and Lorenz, W. (1977). Blood and blood substitutes. *Br. J. Anaesthesiol.* **49**, 681.

Dorner, I., Moore, J.A., Collins, J.A., Sherman, L.A., and Chaplin, H. (1974). Efficacy of buffy coat – poor red cell suspensions prepared by sedimentation in hydroxyethyl starch. *Transfusion* **14**, 510.

—— —— —— —— —— (1975). Efficacy of leukocyte – poor red blood cell suspensions prepared by sedimentation in hydroxyethyl starch. *Transfusion* **15**, 439.

Drescher, W.P., Shih, N., Hess, K., and Tishkoff, G.N. (1978). Massive extracorporeal blood clotting during discontinuous flow leukapheresis. *Transfusion* **18**, 89.

Dubois, M., Gilles, K.A., Hamilton, J.K., Rebers, P.A., and Smith, F. (1956). Colourimetric method for determination of sugars and related substances. *Analyt. Chem.* **28**, 350.

Dürr, H.K., Bode, C., Krupinski, R., and Bode, J.C. (1978). Comparison between naturally occurring macroamylasemia and macroamylasemia induced by hydroxyethyl starch. *Eur. J. Clin. Invest.* **8**, 189.

Eastlund, D.T., Storm, M.B., Zaretzky, J.H., and Britten, A.F.H. (1980). Centrifugation leukapheresis: Lack of effect of hydroxyethyl starch on monocyte chemotaxis. Presented to *The Joint Meeting of the International Society of Hematology and International Society of Blood Transfusion*, Montreal, 16–22 August.

—— —— —— —— and Butterworth, S.J. (1979). Centrifugation leukapheresis: Lack of effect of hydroxyethyl starch on monocyte chemotaxis and generation of a neutrophil chemoattractant by PHA-stimulated lymphocytes. *Blood* **54** (Suppl. 1), 122a.

Ehlenberger, A.G. and Nussenzweig, V. (1977). The role of membrane receptors for C3b and C3d in phagocytosis. *J. exp. Med.* **145**, 357.

Ehrly, A.M., Landgraf, H., Saeger-Lorenz, K., and Hasse, S. (1979). Verbesserung der Fliesseigenschaften des Blutes nach Infusion von niedermolekularer Hydroxyaethylstaerke bei gesunden Probanden. *Infusionsther. Klin. Ernaehr.* **6**, 331.

Eisele, R., Birnbaum, D., Büscher, D., Kotter, D., and Nasseri, M. (1979). Die unterschiedliche Kreislaufwirkung bei schneller Infusion von Hydroxyethylstaerke (HAES), Dextran 60 und Blut beim postoperativen Patienten. *Infusionsther. Klin. Ernaehr.* **6**, 3.

Elmer, O., Goransson, G., Saku, M., and Bengmark, S. (1977). Influence of physiological saline, dextran 70, hydroxyethyl starch, degraded gelatin, and fat emulsion solutions on screen filtration pressure. *Eur. Surg. Res.* **9**, 85.

Enomoto, Y. (1974). Clinical experience with Hespander (surgical patients with gastroduodenal ulcer, gallstone, gastric cancer or breast cancer). *Shinryo-To-Shinyaku* **11**, 2305.

Evangelista, B.S., Green, T.J., Gwilt, D.J., and Brown, T.G. (1969). An experimental study of current therapeutic approaches to endotoxin shock. *Archs Int. Pharmacodyn.* **180**, 57.

Ewald, R.A., Young, A.A., and Crosby, W.H. (1964). Particle formation in dextran solutions. *Milit. Med.* **129**, 952.

Farrales, F.B., Belcher, C.S., and Bayer, W.L. (1975). Effect of hydroxyethyl starch on platelet function following granulocyte collection. *Transfusion* **15**, 508.

—— —— —— and Summers, T. (1977). Effect of hydroxyethyl starch on platelet function following granulocyte collection using the continuous flow cell separator. *Transfusion* **17**, 635.

Farrow, S.P., Hall, M., and Ricketts, C.R. (1970). Changes in the molecular composition of circulating hydroxyethyl starch. *Br. J. Pharmacol.* **38**, 725.

Feliu, E., Woessner, S., Matutes, E., Cardellach, F., Granena, A., Marin, P., Puig, L., Gelabert, A., Vives Corrons, J.L., Monteserrat, E., and Rozman, C. (1980). Study of granulocytes (GN) collected by three methods of leukapheresis (LP). Presented to *The 18th Congress of the International Society of Haematology*, Montreal, Canada, 16–22 August.

Fiala, J. (1979). Use of hydroxyethyl starch in blood transfusion and hematology. *Vnitr Lek* **25**, 897.

Forest, R., Cousineau, L., and Methot, M. (1972). Differential survivals of human erythrocytes frozen in hydroxyethyl starch (HES) by varying extracellular composition as well as RBC population. *Cryobiology* **9**, 317.

Franke, H., Sobotta, E.E., Witzki, G., and Unsicker, K. (1975). Function and morphology of isolated rat kidney following cellfree perfusion with various plasma expanders. *Anaesthesist* **24**, 231.

Freireich, E.J., Hester, J.P., and McCredie, K.B. (1978). Prevention of side effects with the continuous flow blood cell separator. In *Cell-separation and cryobiology* (ed. H. Rainer *et al.*). Schattauer, Stuttgart.

French, D. (1972). Action of amylases as influenced by irregularities in the amylose chain. Presented to *The VI International Symposium on Carbohydrate Chemistry*, Madison, Wisconsin.

—— (1973). Chemical and physical properties of starch. *J. Anim. Sci.* **37**, 1048.

—— (1975). Chemistry and biochemistry of starch. In *Biochemistry of starch* in Biochemistry Series One, (ed. W.J. Whelan) Vol. 5, Butterworths, London.

—— Chan, Y., and England, B. (1974). Mechanism of alpha-amylase action on hydroxyethyl amylose and inhibition by lactones. *Fedn Proc. Fedn Am. Socs exp. Biol.* **33**, 1313.

French, J.E. (1980). Leukapheresis practice in the U.S.: Results of a survey of registered establishments. Presented to *The Conference of Leukapheresis Donor Safety*, Bethesda, Maryland, 8 October, 1980.

—— Jemionek, J.F., and Contreras, T.J. (1980). Liquid and cryopreservation effects on in vitro function of dog granulocyte concentrates prepared for transfusion. *Cryobiology* **17**, 252.

Fudeda, H. (1976). Clinical experience with Hespander. *Shinryo-To-Shinyaku* **13**, 1267.

Fujii, A. (1976). Clinical application of hydroxyethyl starch (HES) – retrospective study of 100 cases. *Geka Shinryo* **18**, 120.

Fujinaga, M., Isemura, M., Ikenaka, T., and Matsushima, Y. (1968). Studies on the substrate specificity of Taka-amylase A. I. The mode of action on partially O-methylated amylases of Taka-amylase A. *J. Biochem.* **64**, 73.

Fujita, T. *et al.* (1971). Clinical application of hydroxyethyl starch (HES). *Med. Consul. New Rem.* **8**, 801.

Fukuda, Y. (1972). Clinical studies on the plasma-expanding effect of Hespander and HES (6% in saline). *J. New Rem. Clin.* **21**, 1405.

Fukutome, T. (1977). An experimental study on hydroxyethyl starch. Its partition in body water and effects on the body fluid dynamics. *Jap. J. Anesthesiol.* **26**, 872.

Fuller, B.J., Pegg, D.E., Walter, C.A., and Green, C.J. (1977). An isolated rabbit kidney preparation for use in organ preservation research. *J. Surg. Res.* **22**, 128.

Futani, H. (1976). Clinical examination of hydroxyethyl starch (Hespander) in delivery bleeding. Effect on bleeding blood of 500–1000 g. *Shinryo-To-Shinyake* **13**, 1905.

Garzon, A.A., Cheng, C., Lerner, B., Lichtenstein, S., and Karlson, K.E. (1967*a*). Hydroxyethyl starch (HES) and bleeding. An experimental investigation of its effect on hemostasis. *J. Trauma* **7**, 757.

—— Pangan, J., Kornberg, E., Antell, H., Karlson, K.E., and Stuckey, J.H. (1967*b*). Effect of prolonged storage on myocardial contractility and GOT levels. *Circulation* **36** (Suppl. II), 117.

Gaver, K.M. (1950). Uniformity 2-substituted glucopyranose polymers. US Patent 2518 135.

Gessner, P.K., Parke, D.V., and Williams, R.T. (1960). Studies in detoxication. The metabolism of glycols. *Biochem. J.* **74**, 1.

Geyer, R.P. (1978). Substitutes for blood and its components. In *Blood substitutes and plasma expanders* (ed. G.A. Jamieson and T.J. Greenwalt) p. 1. Liss, New York.

—— (1980). Artificial blood substitutes: oxygen-binding iron chelates. Grant No. 5R01-HL 17844-07, National Institutes of Health, USA.

Glasser, L. (1977). Discontinuous flow centrifugation leukapheresis and neutrophil function. *Transfusion* **17**, 513.

—— and Huestis, D.W. (1979). Characteristics of stored granulocytes collected from donors stimulated with dexamethasone. *Transfusion* **19**, 53.

—— —— and Jones, J.F. (1977). Functional capabilities of steroid-recruited neutrophils harvested for clinical transfusion. *New Engl. J. Med.* **297**, 1033.

Gleich, G.J., Pineda, A.A., Solley, G.O., and Taswell, H.F. (1979). Cytapheresis for procurement of eosinophils and for the treatment of diseases associated with eosinophilia. Presented to the Workshop on Therapeutic Plasma and Cytapheresis, Rochester, Minnesota, 25–26 April.

Gofferje, H. and Hosslick, V. (1977). Zur Hyperamylasaemie nach Infusion von Hydroxyaethylstaerke unterschiedlicher Molekulargewichtsverteilungen. *Infusionsther. Klin. Ernaehr* **4**, 141.

Gollub, S. (1965). Effects of hydroxyethyl starch on hemostasis in the experimental animal. In *Proceedings of the Third Conference on Artificial Colloidal Agents*, p. 107. NAS-NRC, Washington.

—— and Schaefer, C. (1968). Structural alteration in canine fibrin produced by colloid plasma expanders. *Surg. Gynecol. Obstet.* **127**, 783.

—— —— and Squitieri, A. (1967). The bleeding tendency associated with plasma expanders. *Surg. Gynecol. Obstet.* **124**, 1203.

—— —— Schechter, D.C., and Vanichanan, C. (1969*a*). A study of safer plasma substitutes. *Surg. Gynecol. Obstet.* **128**, 1235.

—— Schechter, D.C., Hirose, T., and Bailey, C.P. (1969*b*). Use of hydroxyethyl starch solution in extensive surgical operations. *Surg. Gynecol. Obstet.* **128**, 725.

Gore, J.M., Hunt, S.M., Curby, W.A., and Lionetti, F.J. (1974). Cryopreservation of human granulocytes with hydroxyethyl starch. *Cryobiology* **11**, 542.

Goto, Y. (1971). Experimental and clinical study of a plasma expander – hydroxyethyl starch. *Jap. J. Anesthesiol.* **20**, 1114.

—— —— —— —— —— —— Miyano, H., Suzuki, S., and Tsuda, T. (1971). Clinical evaluation of hydroxyethyl starch infusion during operation. *Shinryo-To-Shinyaku* **8**, 793.

Goto, Y. *et al.* (1972). Experimental studies on plasma expanders affected to Schwartzman type tissue reaction. *Jap. J. Anesthesiol.* **21**, 740.

–– (1974). Difficulty in blood typing after infusion of plasma substitutes. Relationship to erythrocyte sedimentation rate and rouleaux-formation. *Jap. J. Surg.* **4**, 216.
–– and Matsumoto, K. (1973). Effect of plasma substitute on blood typing and erythrocyte sedimentation rate. *Saishin Igaku* **28**, 552.
–– and Aochi, O. (1973). The rheological effects of plasma expanders upon red blood cells. *Nagoya Med. J.* **18**, 253.
–– –– (1974). Electrophoresis of red blood cells using U-tube method and influence of plasma substitutes. *Asian med. J.* **17**, 125.
–– and Kimura, Y. (1976). Schwartzman type tissue reaction between antibiotics and colloidal plasma substitutes in rabbits, and delaying effect of hydrocortisone. *Nagoya med. J.* **21**, 229.
–– Kimura, Y., and Matsumoto, K. (1972*a*). The rheological effects on red blood cells by plasma expanders – physicochemical study. *Clin. Physiol.* **2**, 485.
–– –– –– Yamahara, T., Noguchi, H., Aochi, O., and Nagata, M. (1972*b*). Clinical evaluation of Hespander. *Jap. J. Anaesthesiol.* **21**, 849.
Granath, K.A. (1958). Solution properties of branched dextrans. *Colloid Sci.* **13**, 308.
–– and Kvist, B.E. (1967). Molecular weight distribution analysis by gel chromatography on sephadex. *J. Chromat.* **28**, 69.
–– Strömberg, R., and de Belder, A.N. (1969). Studies on hydroxyethyl starch. Fractionation and molecular weight distribution by gel chromatography. *Die Staerke* **21**, 251.
Graubner, M., Waldschmidt, R., Mueller-Eckhardt, C., and Loeffler, H. (1978). Experience in leukapheresis therapy using an intermittent flow blood cell separator (IFC). In *Cell-separation and cryobiology* (ed. H. Rainer). Schattauer, Stuttgart.
Greenwood, C.T. (1968). Starch degrading and synthesizing enzymes: A discussion of their properties and action pattern. *Adv. Carbohydr. Chem.* **23**, 281.
–– and Hourston, D.J. (1967). Physicochemical studies on starches. XXXII. Some physicochemical properties of hydroxyethyl starch used as a volume-extender for blood plasma. *Die Staerke* **19**, 243.
–– and Banks, W. (1968). *Synthetic high polymers.* Oliver and Boyd, Edinburgh.
–– and Milne, E.A. (1968). Alpha-amylases: A review of their structure and properties. *Adv. Carbohydr. Chem.* **23**, 281.
–– Muir, D.D., and Whitcher, H.W. (1975). Novel method for preparation of hydroxyethyl starch for cryoprotection of human red blood cells. *Die Staerke* **27**, 109.
–– –– –– (1977). Hydroxyethyl starch as a cryoprotective agent for human red blood cells – relation between molecular properties and cryoprotective effect. *Die Staerke* **29**, 343.
–– –– –– (1978). Preparation of hydroxyethyl starch. British Patent 1514720.
–– –– Banks, W., Boxall, L.M., and Thewlis, B. (1974). Report to U.S. Office of Naval Research No. N00014-72-C-0258.
Gregersen, M.I., Chien, S., and Usami, S. (1965). Studies on the effects of hydroxyethyl starch and other plasma expanders on the viscosity of human blood at low shear rates. In *Proceedings of the Third Conference on Artificial Colloidal Agents.* p. 29. NAS-NRC, Washington, DC.
Grindon, A.J. and Coleman, K. (1979). Pheresis with the IBM 2997. *Transfusion* **19**, 665.

Groth, C.G. (1966). Effect of infused albumin and Rheomacrodex on factors governing the flow properties of human blood. *Acta chim. scand.* **131**, 290.
Grünert, A., Ahnefeld, F.W., and Dick, W. (1978). Pharmacology and elimination of hydroxyethyl starch. Presented at *The Volume Replacement with Hydroxyethyl Starch Seminar*, Erlangen, West Germany.
Gryszkiewicz, A. (1978). Hydroxyethyl-derivative of starch – a new plasma substitute. *Acta haematol. pol.* **9**, 137.
Gunja-Smith, Z., Marshall, J.J., Mercier, C., Smith, E.E., and Whelan, W.J. (1970). The structure of amylopectin. *Febs Lett.* **12**, 101.
Harke, H., Thoenies, R., Margraf, I., and Momsen, W. (1976). The influence of different plasma substitutes on blood clotting and platelet function during and after operation. *Anaesthesist* **25**, 366.
—— *et al.* (1980). Rheological and clotting investigations after the infusion of HES 200/0.5 and dextran 40. A comparative clinical trial. *Anaesthesist* **29**, 71.
Hartung, H.J., Klose, R., and Lutz, H. (1979). Tierexperimentelle Untersuchung zur Volumenwirksamkeit von Hydroxyaethylstaerke 40 000 beim akuten haemorrhagischen Schock des Hundes. *Infusionsther. Klin. Ernaehr.* **6**, 231.
Hayashi, K. and Higashi, H. (1975). A clinical evaluation of plasma substitutes; comparison between low molecular weight dextran, hydroxyethyl starch and high molecular weight hydroxyethyl starch. *Jap. J. Anesthesiol.* **24**, 853.
Hazi, H., Hatada, A., Fukuchi, H., Morimoto, K., Takaori, M., and Oda, T. (1972). Study of viscosity of red cell suspensions into various infusion solutions. *Jap. J. Anesthesiol.* **21**, 11.
Heidenreich, O., aus der Muhlen, K., and Heintze, K. (1975). The effect of the plasma substitutes hydroxyethyl starch and dextran 60 on the kidney function of dogs in acute haemorrhagic shock. *Anaesthesist* **24**, 239.
Hempel, V., Metzger, G., Unseld, H., and Schorer, R. (1975). The influence of hydroxyethyl starch solutions on circulation and on kidney function in hypovolaemic patients. *Anaesthesist* **24**, 198.
Herzig, R.H., Poplack, D.G., and Yankee, R.A. (1974). Prolonged granulocytopenia from incompatible platelets transfusions. *New Engl. J. Med.* **290**, 1220.
Hester, J.P., McCredie, K.B., and Freireich, E.J. (1975). Effects of leucapheresis on normal donors. In *Leucocytes: separation, collection and transfusion* (ed. J.M. Goldman and R.M. Lowenthal). Academic Press, New York.
—— —— —— Applebaum, F., and Deisseroth, A. (1977). Modification of leukocyte (WBC) granulocyte (PMN) collection: The Aminco disposable bowl. *Blood* **50** Suppl. 1, 307.
—— —— —— Kellogg, R.M., Mulzet, A.P., and Kruger, V.R. (1979). Principles of blood separation and component extraction in a disposable continuous-flow single-stage channel. *Blood* **54**, 254.
Higby, D.J., Mishler, J.M., Rhomberg, W., Nicora, R.W., and Holland, J.F. (1975). The effect of a single or double dose of dexamethasone on granulocyte collection with the continuous flow centrifuge. *Vox Sang.* **28**, 243.
Hjermstead, E.T. (1959). Starch hydroxyethylethers and other starch ethers. In *Industrial gums: polysaccharides and their derivatives* (ed. R.L. Whistler, and J.N. BeMiller). Academic Press, New York.
Höcker, P., Pitterman, E., and Blumauer, H. (1976). The use of hydroxyethyl starch (HES), dextran and prednisolone in granulocyte collection with an Aminco cell separator. Presented to *The 2nd International Symposium on Leucocyte Separation and Transfusion*, London, 11–13 October.
Hodges, K.L., Kester, W.E., Wiederich, D.L., and Grover, J.A. (1979). Determination of alcoxyl substitution in cellulose ethers by zeisel-gas chromatography. *Analyt. Chem.* **51**, 2172.

Hölscher, B. (1972). Comparative toxicity studies on prolonged administration of hydroxyethyl starch and dextran 60. In *Shock, metabolic disorders and therapy* (ed. W.E. Zimmerman, I. Staib, and D. Jacobson). Schattauer, Stuttgart.

—— (1973/1974). Blutvolumenersatz durch Hydroxyaethylstaerke oder Dextran 60. *Infusionsther. Klin. Ernaehr.* **4**, 281.

—— (1975). Langzeitverträglichkeit hochdosierter Infusionen von Hydroxyaethylstaerke und Dextran 60 bei normovolaemischen Kaninchen. *Infusionsther. Klin. Ernaehr.* **2**, 215.

—— and Kagel, S. (1976). Zur Blutregeneration nach einem gleichvolumigen Blutaustausch mit Dextran 60 oder Hydroxyaethylstaerke. *Infusionsther. Klin. Ernaehr.* **3**, 250.

—— —— Edeling, M., and Fahrmeier, G. (1975). Comparative investigations of hydroxyethyl starch and Dextran 60 in isovolaemic haemodilution and haemorrhagic shock of rats. *Anaesthesist* **24**, 215.

Homann, B., Pesold, R., Bulow, H., Rietbrock, J., Hess, J., and Weiss, K.H. (1977). Hydroxyaethylstaerke als Plasmasubstitut bei der transurethralen Prostatektomie (TUR) nach der <<cold-punch>> Methode. *Anaesthesist* **26**, 5.

Horii, D. *et al.* (1971). General pharmacology of hydroxyethyl starch. *Clin. Report* **9**, 1187.

Hoshiai, H. (1976). Clinical experience of Hespander in delivery bleeding (examination of effect to maintain blood pressure in 500–1700 of bleeding). *Sanka-To-Fujinka* **43**, 240.

Huestis, D.W. (1978). Intermittent flow centrifugation: leukocyte production. In *Cell-separation and cryobiology* (ed. H. Ranier *et al.*). Schattauer, Stuttgart.

—— Corrigan, J.J., and Johnson, H.V. (1975*a*). Leukapheresis of a five-year-old girl with chronic granulocytic leukemia. *Transfusion* **15**, 489.

—— Price, M.J., White, R.F., and Goodsite, L.M. (1975*b*). Granulocyte collection with the Haemonetics blood cell separator. In *Leukocytes: separation, collection and transfusion* (ed. J.M. Goldman and R.M. Lowenthal). Academic Press, London.

—— —— —— and Inman, M. (1975*c*). Use of hydroxyethyl starch to improve granulocyte collection in the Latham blood processor. *Transfusion* **15**, 559.

—— —— —— —— (1976). Leukapheresis of patients with chronic granulocytic leukemia (CGL), using the Haemonetics Blood Processor. *Transfusion* **16**, 255.

Hulse, J.D., Stoll, R.G., Yacobi, A., Gupta, S.D., and Lai, C.-M. (1980). Elimination of high molecular weight hydroxyethyl starch in rats. *Res. Commun. Chem. Pathol. Pharmacol.* **29**, 149.

Husemann, E. and Resz, R. (1956). Natural and synthetic amylose: hydroxyethyl amylase. *J. Polymer Sci.* **19**, 389.

—— and Kafka, M. (1960). Über natürliche und synthetische Amylose. XII. Über die Verteilung der Substituenten in wasserlöslichen Amyloseaethern. *Makromolekulare Chem.* **41**, 208.

Iacone, A., DiBartolomeo, P., Fioritoni, G., and Tolontano, G. (1976). Granulocyte collection from normal donors with CFC and FL: Granulocyte function and clinical results. Presented to *The 2nd International Symposium on Leucocyte Separation and Transfusion*, London, 11–13 October.

Iba, N. (1975). Clinical studies on hydroxyethyl starch (Hespander) as a plasma substitute in gynecological operations (in surgical patients with myoma of the uterus). *Sanfujinka-No-Sekai* **27**, 75.

Ikeda, K., Tanaka, S., Horii, D., Arai, Y., Nakajima, K., and Kawada, M. (1971).

Effect of hydroxyethyl starch on circulatory blood volume and extracellular fluid volume at hemorrhage. Comparison with blood restoration or dextran for clinical use. *Jap. J. Anesthesiol.* **20**, 639.

Imazu, S., Yamagiwa, K., Komatsu, T., Saito, F., Noguchi, H., Takumi, Y., and Tsuda, K. (1976). Effect of plasma expander infusion on the volume of extravascular lung water. *Jap. J. Anesthesiol.* **26**, 252.

—— —— —— —— —— —— (1977). Effects of continuous positive pressure ventilation on the volume of extravascular lung water. *Jap. J. Anesthesiol.* **27**, 567.

Inokuchi, K. (1950). Colloid chemical study on sodium alginate as a plasma substitute. VI. Agglomerating action of sodium alginate on red cells. *Memoirs of the Faculty of Science, Kyushu University, Series C* **1**, 167.

Inoshita, K. (1980). The experimental and clinical studies of oxystarch in chronic renal failure. *Nippon Jinzo Gakkai Shi* **22**, 315.

Inoue, S., Nagai, Y., Ueno, K., Ota, T., Koike, S., Tizumi, K., and Miyake, T. (1977). Effect of 6 per cent hydroxyethyl starch (HES) in normal saline on blood coagulation. *Jap. J. Anesthesiol.* **21**, 27.

Irikura, T. (1972). Studies on hydroxyethyl starch solution (Hespander), as a plasma substitute. VIII. Chronic toxicity tests by three months administration in rabbits. *Pharmacometrics* **6**, 1103.

—— and Kudo, Y. (1972*a*). Studies on hydroxyethyl starch (Hespander) as a plasma substitute. V. Effect of Hespander on the mesenteric blood flow. *Pharmacometrics* **6**, 1019.

—— —— (1972*b*). Studies on hydroxyethyl starch as a plasma expander. III. Effects of hydroxyethyl starches with various degrees of substitution on the blood pressure in rats. *Pharmacometrics* **6**, 1549.

—— —— and Hirayama, T. (1972*a*). Studies on hydroxyethyl starch solution (Hespander) as a plasma substitute. IV. Effect on circulating blood volume in hemorrhagic rabbits. *Pharmacokinetics* **6**, 1013.

—— Kato, A., and Hirayama, T. (1972*b*). Studies on hydroxyethyl starch solution (Hespander) as a plasma substitute. XIV. Influence of Hespander on renal function. *Pharmacometrics* **6**, 1387.

—— Ohkubo, H., and Hirayama, T. (1972*c*). Studies on hydroxyethyl starch solution (Hespander) as a plasma substitute. III. Effect of Hespander on bled cats. *Pharmacometrics* **6**, 1007.

—— Okada, K., and Tamada, T. (1972*d*). Studies on hydroxyethyl starch solution (Hespander) as a plasma substitute. X. Influence of Hespander on blood coagulation and fibrinolysis. *Pharmacometrics* **6**, 1417.

—— Kudo, Y., Kato, A., and Hirayama, T. (1972*e*). Studies on hydroxyethyl starch solution (Hespander) as a plasma substitute. I. Effects of Hespander on bled rabbits. *Pharmacometrics* **6**, 985.

—— —— —— —— (1972*f*). Studies on hydroxyethyl starch solution (Hespander) as a plasma substitute. II. Effects of Hespander on bled dogs. *Pharmacokinetics* **6**, 993.

—— Shoji, S., Takita, S., and Shinkawa, H. (1972*g*). Studies on hydroxyethyl starch solution (Hespander) as a plasma substitute. VII. Subacute toxicity tests in rabbits. *Pharmacometrics* **6**, 1089.

—— Hosomi, J., Ishiyama, M., and Suzuki, H. (1972*h*). Studies on hydroxyethyl starch (Hespander) as a plasma substitute. IX. Teratogenic studies in mice and rabbits. *Pharmacometrics* **6**, 1119.

—— Tamada, T., Kojima, E., and Kanada, K. (1972*i*). Studies on hydroxyethyl starch solution (Hespander) as a plasma substitute. XI. No anaphylactoid reactions to Hespander. *Pharmacometrics* **6**, 1141.

— — — — (1972*j*). Studies on hydroxyethyl starch solution (Hespander) as a plasma substitute. XII. No antigenicity of Hespander. *Pharmacometrics* **6**, 1149.

— — Ishida, R., Okada, K., and Kanada, K. (1972*k*). Studies on hydroxyethyl starch solution (Hespander) as a plasma substitute. XVIII. Fate of hydroxyethyl starch after intravenous administration of Hespander to rabbits. *Pharmacokinetics* **6**, 1429.

— — — — and Kudo, Y. (1972*l*). Studies on hydroxyethyl starch as a plasma expander. IV. Subacute toxicity tests on high molecular weight hydroxyethyl starch. *Pharmacometrics* **6**, 1557.

— — Kudo, Y., and Ohkubo, S. (1972*m*). Studies on hydroxyethyl starch solution (Hespander) as a plasma substitute. XVII. General pharmacological effects of Hespander. *Pharmacometrics* **6**, 1417.

— — — — Okada, T., and Kanada, K. (1972*n*). Studies on hydroxyethyl starch solution (Hespander) as a plasma substitute. VI. Acute toxicity tests in mice, rats and rabbits. *Pharmacometrics* **6**, 1023.

— Ohkubo, H., Hirayama, T., Yamauchi, M., Imai, S., and Ohkubo, S. (1972*o*). Studies on hydroxyethyl starch solution (Hespander) as a plasma substitute. XV. General pharmacological effects of Hespander. *Pharmacometrics* **6**, 1391.

— Mizuochi, K., Arashima, H., Takita, S., Okada, K., Ishida, R., and Tamada, T. (1972*p*). Studies on hydroxyethyl starch solution (Hespander) as a plasma substitute. XIII. Histological examination on storage and excretion of polysaccharides. *Pharmacometrics* **6**, 1347.

— Higo, K., Maeda, A., Kato, A., Naruke, T., Kasai, S., Akiyama, Y., and Hashimoto, M. (1972*q*). Studies on hydroxyethyl starch solution (Hespander) as a plasma substitute. XVI. General pharmacological effects of Hespander. *Pharmacometrics* **6**, 1409.

Irving, M.H. and Rushman, G.B. (1971). Parenteral nutrition on the surgical patient. *Anaesthesist* **26**, 450.

Isa, T. *et al.* (1971). Study on hydroxyethyl starch. I. Circulatory effect, histological findings and clinical application. *Med. Consul. New Rem.* **8**, 13.

Ishii, S. *et al.* (1971). Clinical experience on hydroxyethyl starch (HES). *Med. Consul. New Rem.* **8**, 827.

Isikura, H. (1976). Clinical experience of Hespander in surgical patients with old ages. *J. New Rem. Clin.* **25**, 1679.

Ito, K. (1972). Effect of hydroxyethyl starch as a plasma expander – clinical evaluation in neurolept anaesthesia. *Med. Consul. New Rem.* **9**, 1667.

Ivankovic, S. and Bülow, I. (1975). On the lacking teratogenic effect of the plasma expander hydroxyethyl starch in the rat and mouse. *Anaesthesist* **24**, 244.

Iwatsuki, K. (1977). Clinical experience of Hespander in ruptured brain aneurysm. *Med. Consul. New Rem.* **14**, 1027.

Janes, A.W., Mishler, J.M., and Lowes, B. (1977). Serial infusion effects of hydroxyethyl starch on ESR, bloodtyping and crossmatching and serum amylase levels. *Vox Sang.* **32**, 131.

Jesch, F., Klövekorn, W.P., Sunder-Plassmann, L., Seifert, J., and Messmer, K. (1975). Hydroxyaethylstaerke als Plasmaersatzmittel: Untersuchungen mit isovolaemischer Haemodilution. *Anaesthesist* **24**, 202.

— Hübner, G., Zumtobel, V., Zimmermann, M., and Messmer, K. (1979). Hydroxyaethylstaerke (HAES 450/0.7) in Plasma und Leber. Konzentrationsverlauf und histologische Veränderungen beim Menschen. *Infusionsther. Klin. Ernaehr.* **6**, 112.

Jeske, A.H., Fonteles, M.C., and Karow, A.M. (1974*a*). Effects of hydroxyethyl starch (HES) and dextran 70 in isolated rabbit kidney. *Cryobiology* **11**, 569.
— — — (1974*b*). Functional effects of nonprotein colloids in the isolated, perfused rabbit kidney. *J. Surg. Res.* **17**, 125.
Kanai, T. (1975). Effects of hydroxyethyl starch (Hespander) on hepatic blood flowing volume. *Med. Consul. New Rem.* **12**, 2287.
Kanehako, F. (1972). Antigenicity of Hespander. *Skinyaku-To-Rinsho* **21**, 871.
Kaneko, K. (1975). Clincial experience of hydroxyethyl starch (Hespander) – comparison between non- and HES-treated groups. *J. New Rem. Clin.* **24**, 1495.
Karlson, K.E., Garzon, A.A., Shaftan, G.W., and Chu, C.J. (1967). Increased blood loss associated with administration of certain plasma expanders. Dextran 75, Dextran 40 and hydroxyethyl starch. *Surgery* **62**, 670.
Katsuya, H., Ohtsu, H., Inove, K., Isa, T., and Morioka, T. (1973). Effects of acute hemorrhage and rapid infusion of colloid solution on the pulmonary shunt ratio. *Anesth. Analg.* **52**, 355.
Katz, A.J., Genco, P.V., and Kiraly, T. (1978). Separation of platelets after discontinuous flow leukapheresis. *Transfusion* **18**, 635.
Kawasaki, T., Nakamura, K., Taninaka, K., and Asada, S. (1971). Experience in the use of hydroxyethyl starch (HES) in extracorporeal circulation. *Skinyaku Rinsho* **5**, 1262.
Kawashima, H. (1972). Experimental studies on hydroxyethyl starch as a hemodiluent in extracorporeal circulation. *Geka Shinryo* **14**, 895.
Kerr, R.W. and Faucette, W.A. (1956). Starch ethers by reaction of starch with alkylene oxides. US Patent 2 273 238.
Kesler, C.C. and Hjermstead, E.T. (1950*a*). Starch ethers in original granule form. US Patent 2 516 632.
— — (1950*b*). Cold water swelling starch ethers in original granule form. US Patent 2 516 634.
— — (1950*c*). Starch ethers in original granule form. US Patent 2 516 633.
— — (1958). Hydroxyalkylation of ungelatinised starches and dextrins in aqueous water miscible alcohols. US Patent 2 845 417.
Kilian, J., Spilker, D., and Borst, R. (1975). Wirkung von 6%iger Hydroxyaethylstaerke, 4,5%igem Dextran 60 und 5,5%igem Oxypolygelatin auf Blutvolumen und Kreislauf bei Versuchspersonen. *Aneasthesist* **24**, 193.
Kimura, Y., Inoue, Y., and Takeuchi, Y. (1971). Basic studies on the immunogenicity of hydroxyethyl starch (HES). *Igaku No Ayumi* **78**, 415.
Kimura, K. (1975). Effect of HES (Hespander) on plasma aldosterone level and electrolyte metabolism caused by a modified NLA method and surgical attack. *Med. Consul. New Rem.* **12**, 2103.
— (1976). Clinical results of the use of Hespander, a new plasma expander. *Sanfujinka-No-Sekal* **28**, 155.
Kinoshita, T. (1971). A clinical evaluation of a new plasma expander 'hydroxyethyl starch'. *Wakayama Med. Reports* **15**, 53.
Kirch, W., Köhler, H., and Horstmann, H.J. (1978). Retarded elimination of a high-molecular enzyme-substrate-complex after hydroxyethyl starch infusion. *Archs Toxicol. Suppl.* **1** 335.
Kisker, C.T., Strauss, R.G., Koepke, J.A., Maguire, L.C., and Thompson, J.S. (1979). The effects of combined platelet and leukapheresis on the blood coagulation system. *Transfusion* **19**, 172.
Kitamura, Y. (1972). Blood level of expanders and erythrocyte sedimentation rate. *Jap. J. Anesthesiol.* **21**, 13.

–– Yamada, A., Sha, N., Hamai, R., Nishimura, K., and Fujimori, M. (1972*a*). A clinical study of hydroxyethyl starch. *Osaka City Med. J.* **18**, 21.

–– –– –– –– –– –– (1972*b*). Studies on Hespander (hydroxyethyl starch). *Jap. J. Anesthesiol.* **21**, 1121.

Kitsutaka, S. (1976). Clinical experience of Hespander in 20 cases with gynecological operation. *Kiso-To-Rinsho* **10**, 2788.

Kleine, N. (1975). The influence of the plasma-expander hydroxyethyl starch on the serological blood grouping tests as compared to dextran 60 and whole blood. *Anaesthesist* **24**, 225.

Knorpp, C.T., Spencer, H.H., Starkweather, W.H., and Gikas, P. (1968). Hydroxyethyl starch as an extracellular cryophylactic agent. *Cryobiology* **4**, 255.

–– –– –– and Hillel, H. (1969). Preliminary results on survival of HES-protected frozen-thawed blood in humans. *Cryobiology* **6**, 268.

–– –– –– and Weatherbee, L. (1971*a*). The preservation of erythrocytes at liquid nitrogen temperatures with hydroxyethyl starch: The removal of hydroxyethyl starch from erythrocytes after thawing. *Cryobiology* **8**, 511.

–– –– –– –– (1971*b*). Survival of frozen erythrocytes protected by hydroxyethyl starch in primates after post-thaw washing. *Cryobiology* **8**, 392.

–– Merchant, W.R., Gikas, P.W., Spencer, H.H., and Thompson, N.W. (1967*a*). In vitro studies on cryoprotective ability of hydroxyethyl starch in erythrocyte preservation. *Cryobiology* **3**, 370.

–– –– –– –– –– (1967*b*). Hydroxyethyl starch: Extracellular cryophylactic agent for erythrocytes. *Science N.Y.* **157**, 1312.

Kobayashi, K., Takagi, Y., Tanifugi, Y., and Sugahara, Y. (1971). Clinical study on hydroxyethyl starch (HES). *Jap. J. Anesthesiol.* **20**, 742.

Koch, K., Jackson, D., Thompson, W.L., and Schmiedl, M. (1978). Total cerebral ischemia (TCI) – Therapy with hydroxyethyl starch (HES) and dopamine. *Clin. Res.* **26**, A290.

Koepke, J.A., Parks, W.M., Goeken, J.A., Klee, G.G., and Strauss, R.G. (1981). The safety of weekly plateletpheresis: Effect upon the donors' lymphocyte population. *Transfusion* **21**, 59.

Köhler, H. (1978). Nebenwirkungen von kolloidalen Plasmaersatzmitteln. *Intensivbehandlung* **3**, 138.

–– Kirch, W., and Horstmann, H.J. (1977*a*). Hydroxyethyl starch-induced macroamylasemia. *Int. J. Clin. Pharmacol. Biopharm.* **15**, 428.

–– –– –– (1977*b*). Die Bildung hochmolekularer Komplexe aus Serumamylase und kolloidalen Plasmaersatzmitteln. *Anaesthesist* **26**, 623.

–– –– and Pitz, H. (1977*c*). Volumenzweiteffekt nach einmaliger Infusion von Hydroxyaethylstaerke. *Verh. Dtsch. Ges. Inn. Med.* **83**, 1756.

–– –– –– (1978*a*). Second increase in plasma volume after single infusion of hydroxyethyl starch. *Klin. Wochenshr.* **56**, 977.

–– –– Vogt, J., and Höffler, D. (1977*d*). Pharmakokinetik von Hydroxyaethylstaerke bei Niereninsuffizienz. *Verh. Dtsch. Ges. Inn. Med.* **83**, 1676.

–– –– Klein, H., and Distler, A. (1978*b*). Die Volumenwirkung von 6% Hydroxyaethylstaerke 450/0.7, 10% Dextran 40 und 3.5% isozyanatvernetzter Gelatine bei Patienten mit terminaler Niereninsuffizienz. *Anaesthesist* **27**, 421.

–– –– Weihrauch, T.R., Prellwitz, W., and Horstmann, H.J. (1977*e*). Macroamylasemia after treatment with hydroxyethyl starch. *Eur. J. Clin. Invest.* **7**, 205.

–– –– Roloff, B., Weihrauch, T.R., Prellwitz, W., and Höffler, D. (1976). Beeinflussung der Serumamylase durch kolliodale Volumenersatzmittel. *Verh. Dtsch. Ges. Inn. Med.* **82**, 968.

Konertz, W., Bischoff, K., Hilsentiz, G., Romeike, J., Westernhagen, T.V., and Bernhard, A. (1976). Stoffwechsel und Organfunktionen bei hypothermer, blutfreier Ganzkörperfusion mit Hydroxyaethylstaerke (HAES) im Tierexperiment. *Langenbecks Arch. Chir.*, Suppl. 48.

Kono, K. *et al.* (1971). Experience on application of HES against hamorrhage in operation – a comparison with Dextran 70 or whole blood. *Shinyaku Rinsho* **5**, 1215.

—— *et al.* (1972). Experimental and clinical investigations on the determination method of hydroxyethyl starch concentration in blood. *Jap. J. Anesthesiol.* **21**, 860.

Koop, C.E. and Bullit, L. (1945). Gelatin as a plasma substitute. The effect of gelatin infusion on the subsequent typing and crossmatching of blood, with a method of eliminating the phenomenon of pseudo-agglutination. *Am. J. Med. Sci.* **209**, 28.

Körber, C. and Scheiwe, M.W. (1977). Effect of hydroxyethyl starch on NaCl-H_2O phase-diagram and its influence on freezing process of cells. *Cryobiology* **14**, 705.

—— —— (1980). The cryoprotective properties of hydroxyethyl starch investigated by means of differential thermal analysis. *Cryobiology* **17**, 54.

Kori-Lindner, C. and Hubert, H. (1978). Klinische Erfahrungen mit einer 6%-igen Hydroxyaethylstaerke – 40 000 – Loesung: Kreislaufparemeter, Vertraeglichkeit. *Infusionsther. Klin. Ernaehr.* **5**, 3.

Kosaka, F. (1975). Preventive effect on hypotension at spinal anaesthesia application of Hespander as a plasma expander. *Rinsho-To-Kenkyn* **52**, 867.

Kościelak, J., Gryszkiewicz, A., Schier, J., and Góralski, S. (1977). Fate of ^{14}C-labeled hydroxyethyl starch in mice. *Acta haematol. pol.* **8**, 303.

Kotelba-Witkowska, B. and Gryszkiewicz, A. (1975). Use of hydroxyethyl starch as a cryoprotective medium during platelet storage at low temperatures. *Acta haematol. pol.* **6**, 293.

Kox, W., Busch, H., and Kiehl, R. (1978). Der Einfluss verschiedener Plasmaexpander auf blutgruppenserologische Untersuchungsergebnisse. *Infusionsther. Klin. Ernaehr.* **5**, 337.

—— and Howekamp, Ch. (1979). Der Einfluss verschiedener Plasmaersatzmittel auf blutgruppenserologische Untersuchungen. *Infusionther. Klin. Ernaehr.* **6**, 341.

Koyama, K. and Yamauchi, H. (1972). Experience with Hespander. I. Experimental study. *Med. Consul. New Rem.* **9**, 1245.

Kraatz, J., van Ackern, K., Glocke, H., Martin, E., Peter, K., and Schmitz, E. (1975). Circulatory changes during preoperative isovolaemic haemodilution with a mixed solution of hydroxyethyl starch and 5%human albumin. *Anaesthesist* **24**, 210.

Kuba, J. (1972). Maintaining effect of Hespander on circulating blood volume during operation. *Shinyaka-To-Rinsho* **21**, 1399.

Kubota, M. (1972). Rheological studies on Hespander in massive hemorrhage. *Shinyaka-To-Rinsho* **21**, 1393.

—— (1975*a*). Studies on Hespander as hemodiluent at extracorporeal circulation. *Geka Shinryo* **17**, 790.

—— (1975*b*). Clinical experience of hydroxyethyl starch solution (Hespander) in the neurosurgical field. *Med. Consul. New Rem.* **12**, 631.

Kudo, K. *et al.* (1971). Clinical use of hydroxyethyl starch as plasma expander in surgical operation. *Shinyaku Rinsho* **5**, 1251.

Lamkc, L.-O. and Liljedahl, S.-O. (1977). Plasma volume changes after infusion of various plasma expanders. *Resuscitation* **5**, 93.

Lancet (1971). Plasma-volume replacement with hydroxyethyl starch. (Editorial.) **ii**, 147.
Landis, E.M. and Pappenheimer, J.R. (1963). Exchange of substances through the capillary walls. In *Handbook of physiology*, Section 2, Circulation, Vol. 11, p. 1017. Waverly Press, Baltimore.
Lane, T.A. (1980). Continuous-flow leukapheresis for rapid cyto-reduction in leukemia. *Transfusion* **20**, 455.
Launer, H.F. and Tominatsu, Y. (1952). Rapid method for moisture in fruits and vegetables by oxidation with dichromate. *Food Technol.* **59**, 235.
—— —— (1953). Rapid accurate determination of carbohydrates and other substances with dichromate heat-of-dilution method. *Analyt. Chem.* **25**, 1767.
Lazrove, S., Waxman, K., Shippy, C., and Shoemaker, W.C. (1980). Hemodynamic, blood volume, and oxygen transport responses to albumin and hydroxyethyl starch infusion in critically ill postoperative patients. *Crit. Care Med.* **8**, 302.
Lee, W.H. and Clowes, G.H.A. (1965). A comparison of the acute hematologic and hemodynamic effects of dextran and hydroxyethyl starch infusions following thermal burn. In *Proceedings of the Third Conference on Artificial Colloidal Agents*, p. 131. NAS-NRC, Washington, DC.
—— Vujovic, V., and Clowes, G.H.A. (1965). Changes of hemodynamics and blood coagulation in burns. *Fedn Proc. Fedn Am. Socs exp. Biol.* **24**, 340.
—— Rubin, J.W., and Huggins, M.P. (1975). Clinical evaluation of priming solutions for pump oxygenator perfusion. *Ann. Thorac. Surg.* **19**, 529.
—— Cooper, N., Weidner, M.G., and Murner, E.S. (1968). Clinical evaluation of a new plasma expander, hydroxyethyl starch. *J. Trauma* **8**, 381.
—— Najib, A., Weidner, M., Clowes, G.H.A., Murner, E.S., and Vujovic, V. (1967). The significance of apparent blood viscosity in circulatory hemodynamic behavior. In *Hemorheology*. Pergamon Press, Oxford.
Lee-Benner, L. and Walton, R.P. (1965). Comparative stability of hydroxyethyl starch and dextran solutions in storage. In *Proceedings of the Third Conference on Artificial Colloidal Agents* p. 54. NAS-NRC, Washington, DC.
Lenzhofer, R., Moser, K., Piller, G., and Rainer, H. (1978). Lymphocyte preparation for immunotherapy. In *Cell-separation and cryobiology* (ed. H. Rainer, *et al.*). Schattauer, Stuttgart.
Lewis, J.H., Szeto, I.L.F., Bayer, W.L., Takaori, M., and Safar, P. (1966). Severe hemodilution with hydroxyethyl starch and dextrans. *Archs. Surg.* **93**, 941.
Lichtenfeld, K.M. and Schiffer, C.A. (1979). The effect of dexamethasone on platelet function. *Transfusion* **19**, 169.
Lindblad, G. (1970). The toxicity of hydroxyethyl starch: Investigation in mice, rabbits, and dogs. *Proc. Eur. Soc. Study Drug Toxicol.* **11**, 128.
—— and Falk, J. (1976). Konzentrationsverlauf von Hydroxyaethylstaerke und Dextran in Serum und Lebergewebe von Kaninchen und die histopathologischen Folgen der Speicherung von Hydroxyaethylstaerke. *Infusionsther. Klin. Ernaehr.* **3**, 301.
Lionetti, F.J. and Hunt, S.M. (1974). Preservation of human red cells in liquid nitrogen with hydroxyethyl starch. *Cryobiology* **11**, 537.
—— —— (1975). Cryopreservation of human red-cells in liquid nitrogen with hydroxyethyl starch. *Cryobiology* **12**, 110.
—— and Callahan, A.B. (1979). Cryogenic preservation of full units of human RBCs with HES 450/0.70. *Vox Sang.* **37**, 364.
—— —— and Lin, P.S. (1975*a*). Cryopreservation of full-units of human erthyrocytes with hydroxyethl starch (HES) in liquid-nitrogen. *Cryobiology* **12**, 561.

—— —— —— (1976). Improved method for the cryopreservation of human red cells in liquid nitrogen with hydroxyethyl starch. *Cryobiology* **13**, 489.

—— —— Curby, W.A., and Gore, J.M. (1974). Cryopreservation of human granulocytes. *Transfusion* **14**, 510.

—— —— —— —— (1975*b*). Cryopreservation of human granulocytes. *Cryobiology* **12**, 181.

—— —— Mattaliano, R.J., and Valeri, C.R. (1978). In vitro studies of cryopreserved baboon granulocytes. *Transfusion* **18**, 685.

—— Luscinskas, F.W., Hunt, S.M., Valeri, C.R., and Callahan, A.B. (1980). Factors affecting the stability of cryogenically preserved granulocytes. *Cryobiology* **17**, 297.

Lorenz, W. (1975). Histamine release in man. *Agents Actions* **5**, 402.

—— and Doenicke, A. (1978). Histamine release in clinical conditions. *Mt. Sinai J. Med. N.Y.* **45**, 357.

—— —— Reimann, H.J., Schmal, A., Schwarz, B., and Dormann, P. (1978). Anaphylactoid reactions and histamine release by plasma substitutes: A randomized controlled trial in human subjects and in dogs. *Agents Actions* **8**, 397.

—— —— Freund, M., Schmal, A., Dormann, P., Praetorius, B., and Schurk-Bulrich, M. (1975). Plasmahistaminspiegel beim Menschen nach rascher Infusion von Hydroxyaethylstaerke: ein Beitrag zur Frage Allergischer Reaktionen nach Gabe eines neuen Plasmasubstituts. *Anaesthesist* **24**, 228.

Lortz, H.J. (1956). Determination of hydroxyethyl groups in low-substituted starch ethers. *Analyt. Chem.* **28**, 892.

Lott, C.E. and Brobst, K.M. (1966). Gas chromatographic investigation of hydroxyethyl amylose hydrolyzates. *Analyt. Chem.* **38**, 1767.

McCredie, K.B. and Freireich, E.J. (1971). Increased granulocyte collection from normal donors with increased granulocyte recovery following transfusion. *Proc. Am. Ass. Cancer Res.* **12**, 58.

—— —— Hester, J.P., and Vallejos, C.S. (1973). Leukocyte transfusion therapy for patients with host-defense failure. *Transplant. Proc.* **5**, 1285.

—— —— —— —— (1974). Increased granulocyte collection with the blood cell separator and the addition of etiocholanolone and hydroxyethyl starch. *Transfusion* **14**, 357.

McCullough, J., Weiblen, B.J., Deinard, A.R., Boen, J., Fortuny, I.E., and Quie, P.G. (1976). In vitro function and post-transfusion survival of granulocytes collected by continuous-flow centrifugation and by filtration leukapheresis. *Blood* **48**, 315.

McGann, L.E. (1978). Differing actions of penetrating and nonpenetrating cyroprotective agents. *Cryobiology* **15**, 382.

Maeda, K. (1973). Clinical trial of hydroxyethyl starch (HES) – surgical cases in obstetrics and orthopedic fields. *Shikoku Igaku Zassi* **29**, 371.

Maguire, L.C., Strauss, R.G., Henriksen, R.A., and Goedken, M.M. (1978). The effect of intermittent-flow centrifugation plateletpheresis (PP) or leukapheresis (LP) on platelet function in donors and units. *Blood* **52** (Suppl 1), 300.

—— —— and Koepke, J.A. (1979*a*). Elimination of hydroxyethyl starch from donor blood after single and multiple leukapheresis. *Transfusion* **19**, 668.

—— —— —— and Henriksen, R.A. (1979*b*). Effects of intermittent-low centrifugation plateletpheresis or leukapheresis on donor platelet functions. *Transfusion* **19**, 660.

—— —— —— —— Stein, M.N., and Thompson, J.S. (1979*c*). Properties of platelets prepared for transfusion by intermittent-flow centrifugation platelet and leukapheresis. *Blood* **54** Suppl. 1, 126a.

—— —— —— —— Goedken, M.M. Echternacht, B., and Thompson, J.S. (1980*a*). Platelet function in donors undergoing intermittent-flow centrifugation plateletpheresis or leukapheresis. *Transfusion* **20**, 549.

—— —— Koepke, J.A., Bowman, R.J., Zelenski, K.R., Lambert, R.M., Hulse, J.D., and Atnip, A.K. (1981). The elimination of hydroxyethyl starch from the blood of donors experiencing single or multiple intermittent-flow centrifugation leukapheresis. *Transfusion* **21**, 347.

—— Henriksen, R.A., Strauss, R.G., Stein, M.N., Goedken, M.M., Echternacht, B., Koepke, J.A., and Thompson, J.S. (1980*b*). Function and morphology of platelets produced for transfusion by intermittent-flow centrifugation plateletpheresis or combined platelet-leukapheresis. *Transfusion* (in press).

Martin, E., Armbruster, I., Fischer, E., Graatz, J., Kersting, H., Oberst, R., and Peter, K. (1976). Gerinnungsveraenderungen bei Anwendung verschiedener Dilutionsloesungen bei praeoperativer isovolaemischer Haemodilution. *Anaesthesist* **25**, 181.

Matsuda, T. and Murakami, M. (1972). Effects of hydroxyethyl starch on blood coagulation and fibrinolysis. I. In vitro studies, *Jap. J. Clin. Haematol.* **13**, 831.

—— —— and Hashizume, K. (1972). Effects of hydroxyethyl starch on blood coagulation and fibrinolysis. II. In vivo studies. *Jap. J. Clin. Haematol.* **13**, 934.

Matsui, K. (1974). Clinical experience with hydroxyethyl starch solution (Hespander) during extracorporeal circulation (with regard to safety). *Rinsho-To-Kenkyu* **51**, 1361.

Matsukawa, S. (1976). Clinical experience of hydroxyethyl starch (Hespander) in patients with extensive burns. *J. New Rem. Clin.* *25, 257.*

Matsumoto, I., Hanafusa, M., Kumazawa, T., Tsunoda, Y., and Ikezono, E. (1977). Experimental study on the effects of plasma substitutes (6% Hespander, 3% Dextran 40) on blood coagulation. *Jap. J. Anesthesiol.* **26**, 59.

—— *et al.* (1974). Experimental study on the effects of plasma substitutes (HES, Dextran 70) on blood coagulation. *Jap. J. Anesthesiol.* **23**, 1323.

Matsumoto, K. (1974). Influence of various kinds of artificial colloids on electric charge of erythrocytes, with special regard to hydroxyethyl starches. *Rinsho-To-Kenkyu* **51**, 844.

Matsumoto, T. *et al.* (1971). Whether or not hydroxyethyl starch has antigenecity. *Shinyaku Rinsho* **5**, 1183.

Matsuoka, S. (1975). Effect of HES on hypotension after photographing of vessels, comparison with 5% glucose. *Med. Consul. New Rem.* **12**, 1899.

—— (1977). Effect of HES on hypotension after photographing of vessels in 300 patients. Comparison on blood pressure change by 5% glucose, Ringer's solution and HES. *Med. Consul. New Rem.* **14**, 1189.

Matsuura, Y., Tsubokura, T., and Tamura, M. (1973). Comparative study on the influences of plasma substances on the blood coagulation. *Jap. J. Surg.* **3**, 80.

Maurer, P.H. (1965). Immunogenicity studies with hydroxyethyl starch. In *Proceedings of the Third Conference on Artificial Colloidal Agents* p. 26. NAS-NRC, Washington, D.C.

—— and Berardinelli, B. (1968). Immunological studies with hyroxyethyl starch (HES); a proposed plasma expander. *Transfusion* **8**, 265.

Merkus, H.G., Mourits, J.W., Degalan, L., and Dejong, W.A. (1977). Substitution distribution in hydroxyethyl starch. *Die Staerke* **29**, 406.

Meryman, H.T. (1972). A possible mechanism of cryoprotection by extracellular agents. *Cryobiology* **9**, 321.

— and Hornblower, M. (1978). Advances in red cell freezing. *Transfusion* **18**, 632.

Messmer, K. and Jesch, F. (1978). Volumenersatz und Hämodilution durch Hydroxyaethylstaerke. *Infusionsther. Klin. Ernaehr.* **5**, 169.

— Ring, J., Hedin, H., Richter, W., and Seemann, C. (1978). Nebenwirkungen bei der Anwendung kolloidaler Infusionslösungen. *Zentralbl. Chir.* **103**, 978.

Metcalf, W., Papadopoulos, A., Tufaro, R., and Barth, A. (1970). A clinical physiologic study of hydroxyethyl starch. *Surg. Gynecol. Obstet.* **131**, 255.

Meyer, K.H., Berfeld, P., and Hohenemser, W. (1941). Acetates and nitrates of amylose and amylopectin. *Helv. chim. Acta* **23**, 885.

Mima, Y. and Gokoyama, K. (1970). Hydroxyethyl starch. Japanese Patent No. 70 06 556.

Mishler, J.M. (1975). Hydroxyethyl starch as an experimental adjunct to leukocyte separation by centrifugal means: Review of safety and efficacy. *Transfusion* **15**, 449.

— (1977*a*). Enhancement of phagocytosis by human neutrophils incubated with DXM, hydroxyethylated amylopectin and Na_3-citrate. In *Blood leucocytes – function and use in therapy* (ed. C.F. Hogman *et al.*). ISBT Publication, Uppsala.

— (1977*b*). The effects of corticosteroids on mobilization and function of neutrophils. *Exp. Hematol.* **5** (Suppl. 1.), 15.

— (1978*a*). Granulocyte replacement therapy. In-vitro function. D. Phil. thesis, University of Oxford.

— (1978*b*). New dosage regimens for HES during intensive leukapheresis. *Transfusion* **18**, 126.

— (1978*c*). Donor conditioning agents. Usage and effect on in-vitro and in-vivo neutrophil function. In *Cell-separation and cryobiology*. (ed. H. Rainer, *et al.*). Schattauer, Stuttgart.

— (1979*a*). The uptake of ^{3}H-dexamethasone during phagocytosis of *staphylococcus aureus* by human neutrophils. *Haematologia* **12**, 209.

— (1979*b*). Designing specific-acting HES cryoprotectants for the freezing of human RBCs. *Vox Sang.* **37**, 359.

— (1979*c*). The plasma kinetics of hydroxyethyl starch 350/0.60. A potential new adjunct for centrifugal leucapheresis. *Am. J. Hematol.* **7**, 341.

— (1980*a*). The hydroxyethylated amylopectins. Model substances comprising a two-variable system for the design of specific acting volaemic colloids. *Int. J. Clin. Pharmacol. Biochem.* **18**, 67.

— (1980*b*). Pharmakokinetik mittelmolekularer Hydroxyaethylstaerke (HAS 200/0.50). *Infusionsther. Klin. Ernaehr.* **7**, 96.

— (1980*c*). Hydroxyethyl starch. Control of alpha-amylase hydrolysis and its subsequent clinical application. *Acta haematol. pol.* **11**, 113.

— (1980*d*). Clinical pharmacology of hydroxyethyl starches. *Blood Transfusion Immunohematol.* **23**, 283.

— and Beez, M. (1979). Die mathematische Beschreibung der intravasalen Ausscheidung von HAES nach wiederholten Infusionen beim Menschen. *Infusionsther. Klin. Ernaehr.* **6**, 119.

— and Dürr, H.-K. (1979). Macroamylasaemia following the infusion of low molecular weight-hydroxyethyl starch in man. *Eur. Surg. Res.* **11**, 217.

— and Parry, E.S. (1979). Tranfusion of hydroxyethylated amylopectin-protected frozen blood in man. I. Plasma clearance and renal excretion of the cryoprotectant. *Vox Sang.* **36**, 337.

—— and Dürr, G.K.–H. (1980). Macroamylasemia induced by hydroxyethyl starch: Confirmation by gel filtration analysis of serum and urine. *Am. J. clin. Pathol.* **74**, 387.

—— and Williams, D.M. (1980). Alkaline phosphatase as a marker of maturity in human neutrophils. Studies in normals dosed with aetiochanolone and prednisolone. *J. clin. Pathol.* **33**, 555.

—— Moser, A.M., and Carter, J.B. (1976). The safety of dexamethasone and hydroxyethyl starch in the multiply leukapheresed donor. *Transfusion* **16**, 170.

—— Parry, E.S., and Petrie, A. (1978*a*). Plasma clearance and renal excretion of erythrocyte cryoprotectant hydroxyethylated amylopectin. *Br. J. Haematol.* **40**, 231.

—— —— and Borberg, H. (1978*b*). The pharmacokinetics of LMW-HES (a new plasma expander). *Blood* **52**, (Suppl. 1), 300.

—— Ricketts, C.R., and Parkhouse, E.J. (1979*a*). Changes in the molecular composition of circulating hydroxyethyl starch following consecutive daily infusions in man. *Br. J. clin. Pharmacol.* **7**, 505.

—— —— —— (1980*a*). Catabolism of low molecular weight-hydroxyethylated amylopectin in man. III. Further degradation of excreted polymer fragments. *Int. J. clin. Pharmacol. Biopharm.* **18**, 120.

—— —— —— (1980*b*). The post-transfusion survival of HES 450/0.70 in man. A long-term study. *J. clin. Pathol.* **33**, 155.

—— —— —— (1980*c*). Changes in the molecular size distribution and post-transfusion survival of hydroxyethyl starch 350/0.60 as influenced by a lower degree of hydroxyethylation. A study in normal man. *J. clin. Pathol.* **33**, 880.

—— —— —— (1981). Urinary excretion kinetics of hydroxyethyl starch 350/0.60 in normovolaemic man. *J. clin. Pathol.* **34**, 361.

—— Borberg, H., Emerson, P.M., and Gross, R. (1977*a*). Hydroxyethyl starch: An agent for hypovolaemic shock treatment. I. Serum concentrations in normal volunteers following three consecutive daily infusions. *J. Surg. Res.* **23**, 239.

—— —— —— —— (1977*b*). Hydroxyethyl starch: An agent for hypovolaemic shock treatment. II. Urinary excretion in normal volunteers following three consecutive daily infusions. *Br. J. clin. Pharmacol.* **4**, 591.

—— —— Ricketts, C.R., and Parkhouse, E.J. (1978*c*). Circulating hydroxyethyl starch composition in man after large consecutive infusions. *Blood* **52** Suppl. 1, 301.

—— Parry, E.S., Sutherland, B.A., and Bushrod, J.R. (1979*b*). A clinical study of low molecular weight-hydroxyethyl starch (a new plasma expander). *Br. J. clin. Pharmacol.* **7**, 619.

—— Hadlock, D.C., Fortuny, I.E., Nicora, R.W., and McCullough, J. (1974*a*). Increased efficiency of leukocyte collection by the addition of hydroxyethyl starch to the continuous flow centrifuge. *Blood* **44**, 571.

—— Higby, D.J., Rhomberg, W., Nicora, R.W., and Holland, J.F. (1975*a*). Leukapheresis: Increased efficiency of collection by the use of hydroxyethyl starch and dexamathasone. In *Leucocytes: separation, collection and transfusion* (ed. J.M. Goldman and R.M. Lowenthal). Academic Press, London.

—— Ricketts, C.R., Parkhouse, E.J., Borberg, H., and Gross, R. (1979*c*). Catabolism of low molecular weight-hydroxyethylated amylopectin in man. I. Changes in the circulating molecular composition. *J. Lab. clin. Med.* **94**, 841.

—— —— —— —— —— (1980*d*). Catabolism of low molecular weight-hydroxyethylated amylopectin in man. II. Changes in the urinary molecular profiles. *Int. J. Clin. Pharmacol. Biopharm.* **18**, 5.

— — Higby, D.J., Rhomberg, W., Cohen, E., Nicora, R.W., and Holland, J.F. (1974*b*). Hydroxyethyl starch and dexamethasone as an adjunct to leukocyte separation with IBM blood cell separator. *Transfusion* **14**, 352.

— — Nicora, R.W., Yoshitake, T., Oishi, K., Kawasaki, T., and Shimizu, T. (1975*b*). Hemodilution with hydroxyethyl starch during cardiopulmonary bypass: Review of a multi-institutional study. *J. Extra-Corp. Technol.* **7**, 140.

Mitchell, R. (1979). A review of extracellular (non-penetrating) cryoprotection of human RBCs. *Vox Sang.* **37**, 353.

Miwa, K. (1976). Experience of the use of Hespander during operation – comparison of non-transfused group and transfused group. *Kiso-To-Rinsho* **10**, 2741.

Miyata, K. (1975). Clinical examination of 6% hydroxyethyl starch (Hespander) as à hemodiluent during extracorporeal circulation. Comparison with 4% Dextran. *Geka Shinryo* **17**, 679.

Miyazaki, M. (1974). Studies on hydroxyethyl starch as a plasma substitute. I. Clinical examination of Hespander (comparison of HES-treated and blood transfused groups). *J. New Rem. Clin.* **23**, 1903.

Mizrahi, A. and Moore, G.E. (1970). Preliminary results on survival of hematopoietic cell lines in freezing media containing hydroxyethyl starch. *Cryobiology* **6**, 576.

— — — — (1971*a*). Long-term preservation of permanent human hematopoietic cell lines. *J. Med.* **2**, 380.

— — — — (1971*b*). Role of sodium carboxymethyl cellulose and hydroxyethyl starch in hemopoietic cell line cultures. *Appl. Micro.* **21**, 754.

Morgan, P.W. (1946). Determination of ethers and esters of ethylene glycol. *Ind. Eng. Chem. Analyt. Ed.* **18**, 500.

Mori, Y. (1975). Influence of Hespander on blood viscosity. *Doctor Salon* **19**, 367.

Mourits, J.W., Merkus, H.G., and Degalan, L. (1976). Gas chromatographic determination of hydroxyethyl derivatives of glucose. *Analyt. Chem.* **48**, 1557.

Müller, N., Popov-Cenić, S., Kladetzky, R.-G., Hack, G., Lang, U., Safer, A., and Rahlfs, V.W. (1976). Hydroxyaethylstaerke und ihr Einfluss auf die intra-sowie postoperative Haemostase. *Infusionsther. Klin. Ernaehr.* **3**, 305.

— — — — — — — — — — (1977). The effect of hydroxyethyl starch on the intra- and postoperative behaviour of haemostatis. *Bibl. Anat.* **16**, 460.

Munoz, E., Raciti, A., Dove, D.B., Stahl, W.M., and del Guercio, L.R.M. (1980). Effect of hydroxyethyl starch versus albumin on hemodynamic and respiratory function in patients in shock. *Crit. Care Med.* **8**, 255.

Murphy, G.P. (1965*a*). The comparative physical properties and physiological effects of hydroxyethyl starch solution. In *Proceedings of the Third Conference on Artificial Colloidal Agents*. NAS-NRC, Washington, DC, p. 74.

— — (1965*b*). The renal effects of acute hemodilution with hydroxyethyl starch, dextran, or saline. *Surg. Gynecol. Obstet.* **121**, 1325.

— — Demaree, O.E., and Gagnon, J.A. (1965). The renal and systemic effects of hydroxyethyl starch solution infusions. *J. Urol.* **93**, 534.

Murray, G.F., Solanke, T., Thompson, W.L., and Ballinger, W.F. (1965). Hydroxyethyl starch as a plasma expander in hemorrhagic shock. *Surg. Forum* **16**, 34.

Muteki, G. *et al.* (1971). Experience on hydroxyethyl starch (HES). *Med. Consul. New Rem.* **8**, 785.

Nakagawa, J. (1973). Influence of hydroxyethyl starch (Hespander) on maintenance of circulating blood volume with time. *Masui-To-Sosei* **9**, 235.

Nakajo, N. (1972). Clinical use of hydroxyethyl starch solution. *Jap. J. Anesthesiol.* **21**, 138.
Nakanishi, Y. (1972*a*). Changes of distribution of circulating red cells following acute hemodilution with plasma substitute solutions. *Jap. J. Anesthesiol.* **21**, 341.
—— (1972*b*). Changes of distribution of circulating red cells following administration of hydroxyethyl starch and crystalloid solution in hemorrhagic shock. *Jap. J. Anesthesiol.* **21**, 771.
Neely, W.B. (1960). Dextran: Structure and synthesis. *Adv. Carbohydr. Chem.* **15**, 341.
Nilsson, G. and Nilsson, K. (1974). Molecular weight distribution determination of clinical dextran by gel permeation chromatography. *J. Chromat.* **101**, 137.
Nishijima, H. (1976). Clinical studies on hemodilution during extracorporeal circulation. Clinical trial of hydroxyethyl starch solution. *Rinsho-To-Kenkyu* **53**, 528.
Nishimoto, K. (1976). Studies on Hespander in patients with laparotomy – with special regard to colloid osmotic pressure of blood and urine. *Shinryo-To-Shinyaku* **13**, 525.
Nishimura, N. *et al.* (1972). Clinical evaluation of Hespander. *Jap. J. Anesthesiol.* **21**, 635.
Nokanishi, M. (1976). Basic and clinical effect of Hespander on chronic arterio-obliterans in legs – intravenous and intra-arterial injection. *J. New Rem. Clin.* **25**, 1673.
Norrman, B. (1969). Substituent distributions in hydroxyethyl starch and dextran. *Svens Pap T* **72**, 50.
Nozue, G. (1972). Clinical studies on the injection of hydroxyethyl starch (HES). Comparison of HES and dextran in gynecological operations. *Sanfujinka-No-Sekai* **24**, 337.
Oda, A. (1972*a*). Behavior of hydroxyethyl starch in plasma and urine following intravenous infusion to operative patients. *Jap. J. Anesthesiol.* **21**, 747.
—— (1972*b*). Fundamental and clinical studies on determination of blood level of hydroxyethyl starch. *Jap. J. Anesthesiol.* **21**, 860.
Odaka, Y. *et al.* (1971*a*). Chronic toxicity of 6% hydroxyethyl starch in saline in rats. *Shinyaku Rinsho* **5**, 1156.
—— *et al.* (1971*b*). Subacute toxicity of 6% hydroxyethyl starch in saline. *Shinyaku Rinsho* **5**, 1127.
Ogawa, H. (1975). Influence of Hespander on cardiovascular system during operation. *Shinryo-To-Shinyaku* **12**, 637.
Okada, K. *et al.* (1971). Clinical application of hydroxyethyl starch. *Shinyaku Rinsho* **5**, 1240.
Oishi, K. (1971). Experimental and clinical study on hydroxyethyl starch (HES) in exracorporeal circulation. *J. New Rem. Clin.* **20**, 1419.
—— Takagi, H., Koba, I., Ueda, S., Akasu, I., Fujino, S., Koga, A., Kuga, S., Wada, H., and Yamamoto, E. (1972). Experimental and clinical studies of hydroxyethyl starch (HES) in extracorporeal circulation. *Kurume. med. J.* **19**, 187.
Oyama, K. (1972). Trial experience of Hespander. I. Experimental examination. *Shinryo-To-Shinyaku* **9**, 1245.
Ozaki, T., Tada, M., and Irikura, T. (1972). A study on distribution of hydroxyethyl groups in hydroxyethyl starch. *J. Pharm. Soc. Jap.* **92**, 1500.
Paulini, K. and Sonntag. W. (1976). Veraenderungen des RHS der Ratte nach parenteraler Gabe von Dextran ($M_{\bar{w}}$ 40 000) und Hydroxyaethylstaerke

($M_{\bar{w}}$ 40 000). Chemische, licht- und elektronenmikroskopische Untersuchungen. *Infusionsther. Klin. Ernaehr.* **3**, 294.

Pegg, D.E. (1977). The water and cation content of nonmetabolizing perfused rabbit kidneys. *Cryobiology* **14**, 160.

Persidsky, M.D. and Ellett, M. (1971*a*). Hydroxyethyl starch as a cryoprotective agent for nucleated mammalian cells. *Cryobiology* **8**, 586.

—— —— (1971*b*). Cryoprotective potentials of hydroxyethyl starch for nucleated mammalian cells. *Cryobiology* **8**, 377.

Peter, K., Gander, H.P., Lutz, H., Nold, W., and Stosick, V. (1975). Die Beeinflussung der Blutgerinnung durch Hydroxyaethylstaerke. Eine klinische Vergleichsuntersuchung. *Anaesthesist* **24**, 219.

Pierce, J.C., Cobb, G.W., and Hume, D.M. (1971). Relevance of HL-A antigens to acute humoral rejection of multiple renal allotransplants. *New Engl. J. Med.* **285**, 142.

Pineda, A.A., Brzica, S.M., and Taswell, H.F. (1977). Continuous- and semicontinuous-flow blood centrifugation systems: therapeutic applications, with plasma-, platelet-, lympha-, and eosinapheresis. *Transfusion* **17**, 407.

Polesky, H.F., McCullough, J., Helgeson, M.A., and Nelson, C. (1973). Evaluation of methods for the preparation of HL-A antigen-poor blood. *Transfusion* **13**, 383.

Polushima, T.V. (1980). Antishock blood substitute based on hydroxyethyl starch. *Probi Gematol Pereliv Krovi* **25**, 40.

Poon, A. and Wilson, S. (1980). Simple manual method for harvesting granulocytes. *Transfusion* **20**, 71.

Popov-Cenić, S., Müller, N., Kladetsky, R.-G., Hack, G., and Lang, U. (1977). Comparable studies of the behaviour of the plasmic coagulation, the fibrinolysis as well as the platelets under application of dextran or hydroxyethyl starch. *Bibl. Anat.* **16**, 322.

—— —— —— —— —— Safer, A., and Rahlfs, V.W. (1976). Durch Praemedikation, Narkose und Operation bedingte Anderungen des Gerinnungs- und Fibrinolysesystems und der Thrombozyten. Einfluss von Dextran und Hydroxyaethylstaerke (HAES) waehrend und nach Operation. *Anaesthesist* **26**, 77.

Price, T.H. and Dale, D.C. (1978). Neutrophil transfusion: Effect of storage and of collection method on neutrophil blood kinetics. *Blood* **51**, 789.

Richter, W. and de Belder, A.N. (1976). Antibodies against hydroxyethyl starch produced in rabbits by immunization with a protein hydroxyethyl starch conjugate. *Int. Archs Allergy Appl. Immun.* **52**, 307.

—— Hedin, H., and Ring, J. (1977). Immunologische Befunde bei der Infusion kolloidaler Loesungen. *Med. Welt.* **28**, 1717.

—— —— —— and Messmer, K. (1978). Adverse reactions to plasma substitutes: Incidence and pathomechanisms. In *Adverse response to intravenous drugs* (ed. J. Watkins and A.M. Ward). Academic Press, London.

Riddick, J.A., Toops, E.E., Wieman, R.L., and Gundiff, R.H. (1954). Physicochemical characterization of clinical dextran. *Analyt. Chem.* **26**, 1149.

Ring, J. and Messmer, K. (1977*a*). Incidence and severity of anaphylactic reactions to colloid volume substitute. *Lancet* **i**, 466.

—— —— (1977*b*). Infusionstherapie mit kolloidalen Volumenersatzmitteln. *Anaesthesist* **26**, 279.

—— —— Seifert, J., and Brendel, W. (1976). Anaphylactoid reactions due to hydroxyethyl starch infusion. *Eur. Surg. Res.* **8**, 389.

—— Sharkoff, D., and Richter, W. (1980). Intravascular persistence of hydroxyethyl starch (HES) after serial granulocyte collections using HES in man. *Vox Sang.* **39**, 181.

Rittmeyer, P. (1976). Klinische Erfahrungen mit mehr als 4000 Infusionen von Hydroxyaethylstaerke. *Klinikarzt* **5**, 9.

Roberts, M. (1965). The physiochemical characteristics of hydroxyethyl starch solution. In *Proceedings of the Third Conference on Artificial Colloidal Agents*, p. 35. NRC-NAS, Washington, DC.

—— and Pagones, J.N. (1965). Preliminary tests of hydroxyethyl starch solution in animals. In *Proceedings of the Third Conference on Artificial Colloidal Agents*, p. 120. NRC-NAS, Washington, DC.

Robson, D.C. (1970). *Modern methods of blood preservation* (ed. Spielmann and Seidle). Gustav Fischer, Stuttgart.

Robyt, J.F. and French, D. (1970). The action pattern of porcine pancreatic α-amylase in relationship to the substrate binding size of the enzyme. *J. biol. Chem.* **245**, 3917.

Rock, G. and Wise, P. (1978). Plasma expansion during graunulocyte procurement: The accumulative effects of hydroxyethyl starch. *Blood* **52** Suppl. 1, 302.

—— —— (1979). Plasma expansion during granulocyte procurement: Cumulative effects of hydroxyethyl starch. *Blood* **53**, 1156.

—— Blanchette, V., and Wong, S.C. (1980). Combined platelet–granulocyte procurement: The effect of hydroxyethyl starch on platelet function. Presented to *The 18th Congress of the International Society of Haematology*. Montreal, Canada, 16–22 August.

Roe, J.H. (1954). The determination of dextran in blood and urine with antrhone reagent. *J. biol. Chem.* **208**, 889.

Rowe, A.W. and Peterson, J. (1971). Effect of glycerol, HES, and DMSO on functional integrity of human blood platelets before and after freezing. *Cryobiology* **8**, 397.

Roy, A.J., Simmons, W.B., Franklin, A., and Djerassi, I. (1970). Hydroxyethyl starch for separation of normal granulocytes. *Fedn Proc. Fedn Am. Socs exp. Biol.* **29**, 424.

—— —— —— —— (1971). A method for separation of granulocytes from normal human blood using hydroxyethyl starch. *Prep. Biochem.* **1**, 197.

Russell, H.E., Bradham, R.R., and Lee, W.H. (1966). An evaluation of infusion therapy (including dextran) for venous thrombosis. *Circulation* **33**, 839.

Ryan, A.J., Holder, G.M., Mate, C., and Adkins, G.K. (1972). Metabolism and excretion of hydroxyethyl starch in rat. *Xenobiotica* **2**, 141.

Sadota, I. (1974). Process for manufacture of hydroxyethyl starch. Japanese Patent No. 7 405 193.

—— (1975). Process for producing hydroxyethyl starches. British Patent No. 1 395 777.

Safar, P., Takaori, M., Kirimli, B., Kampschulte, S., and Nemoto, E. (1978). Plasma substitutes for resuscitation. In *Blood substitutes and plasma expanders* (ed. G.A. Jamieson and T.J. Greenwalt). Liss, New York.

Saito, K. (1972). Effects of Hespander, a new substitute for circulating blood plasma, upon blood pressure and the equilibrium between acid and bases in blood. *J. New Rem. Clin.* **21**, 1107.

Saito, T., Sakata, S., Okazaki, K., Tomino, T., Ohta, N., and Tonogai, R. (1976). Effects of isovolaemic hemodilution with hydroxyethyl starch on coronary and systemic circulations in dogs. *Jap. J. Anesthesiol.* **25**, 29.

Sakamato, R., Kojima, T., and Yamaguchi, S. (1977). Characterization of hydroxyethyl starch used as a plasma expander. *Kobunsh Roncho* **34**, 275.

Sakauchi, G. (1975). Experience of the use of Hespander during extracorporeal circulation. *J. New Rem. Clin.* **24**, 1911.

Sakio, H. (1976). Experience of the use of Hespander for prevention of hypotension during spinal anesthesia. *Shinryo-To-Shinyaku* **13**, 1011.

Sato, S. (1976). Clinical experience of Hespander (in case of administration to burned patients). *Kiso-To-Rinsho* **10**, 2832.

Satoshi, A. *et al.* (1971). Clinical evaluation of hydroxyethyl starch. *Jap. J. Clin. exp. Med.* **48**, 2643.

Schaefer, U.W. and Beyer, J.H. (1975). Protective effect of hydroxyethyl starch (HES) in cryoconservation of bone-marrow of mouse and of man. *Anaesthesist* **24**, 505.

—— Nowrousian, M.R., Öhl, S., and Schmidt, C.G. (1978). Cryopreservation of bone marrow. In *Cell-separation and cryobiology* (ed. H. Rainer *et al.*). Schattauer, Stuttgart.

Schaefer, C., Squitieri, A., and Gollub, S. (1966). Hemorrhage with use of plasma expanders. *Fedn Proc. Fedn Am. Socs exp. Biol.* **25**, 620.

Scheiwe, M.W. (1972). Test der Gefrierschutzmittel HES, Dextran und Haemaccel. II. Optimale Kühlgeschwindigkeit fuer Erythrozyten. *Biomed. Technik* **22**, 415.

—— and Krause, K.H. (1977). Test der Gefrierschutzmittel HES, Dextran und Haemaccel. I. Erythrozytenvolumen und Haemolyse in hypertonen Kochsalzloesungen. *Biomed. Technik* **22**, 413.

—— and Nick, H.E. (1977). Effect of extracellular and intracellular water distribution on optimal hydroxyethyl starch concentrations for red blood cells at different cooling and thawing rates. *Cryobiology* **14**, 709.

—— —— and Körber, C. (1979). Physical and chemical aspects of cryopreservation of human RBCs with HES 450/0.70. *Vox Sang.* **37**, 354.

Schiffer, C.A., Aisner, J., Schmukler, M., Whitaker, C.L., and Wolff, J.H. (1975). The effect of hydroxyethyl starch on in-vitro platelet and granulocyte function. *Transfusion* **15**, 473.

—— —— Daly, P.A., Schimpff, S.C., and Wiernik, P.H. (1979). Alloimmunization following prophylactic granulocyte transfusion. *Blood* **54**, 766.

Schoch, T.J. (1963). *Second Conference on Artificial Colloidal Agents.* NAS-NRC, Washington, DC.

—— (1965). Preparation and characterization of hydroxyethyl starch. In *Proceedings of the Third Conference on Artificial Colloidal Agents.* p. 6. NAS-NRC, Washington, DC.

Scholander, H. and Myrback, K. (1951). Amylases in milling products – use of hydroxyethyl starch for determination of α-amylase. *Svensk. Chem. Tid.* **63**, 250.

Schöning, B. and Koch, H. (1975). Randomised study on histamine-like side-effects of 5 common plasma substitutes used in orthopedic surgery. *Anaesthesist* **24**, 507.

Seidemann, J. (1979). Hydroxyethyl starch – a new interesting colloidal plasma volume expander. Anaphylactoid reactions. *Zentralbl Chir* **104**, 1457.

Senti, F.R., Hellman, N.N., Ludwig, N.H., Babcock, G.E., Tobin, R., Glass, C.A., and Lomberts, B.L. (1955). Viscosity, sedimentation and light-scattering properties of fractions of an acid hydrolysed dextran. *J. Polymer Sci.* **17**, 527.

Shields, C.E., Adner, M.M., and Eichelberger, J. (1965). Characteristics of hydroxyethyl starch solution in storage and antigenicity and blood coagulation studies. In *Proceedings of the Third Conference of Artificial Colloidal Agents*, p. 16. NAS-NRC, Washington, DC.

Shimizu, K., Yokochi, H., and Fukukei, I. (1971). Clinical trial of hydroxyethyl starch (HES). *Med. Consul. New Rem.* **8**, 809.

Shiraki, S. (1976). A preventive method for hypotension during Cesarean

section by spinal anesthesia. Transfusion of low molecular HES prior to spinal anesthesia. *Sanfujinka-No-Sekai* **28**, 661.
Silk, M.R. (1966). The effect of dextran and hydroxyethyl starch on renal hemodynamics. *J. Trauma* **6**, 617.
Skaer, H.L.B., Franks, F., and Echlin, P. (1978). Non-penetrating polymeric cryofixatives for ultrastructural and analytical studies of biological tissues. *Cryobiology* **15**, 589.
Smith, J.A.R., Norman, J.N., Smith, A., and Smith, G. (1975). Comparison of dextran 70 and hydroxyethyl starch in volume replacement. *Br. J. Surg.* **62**, 666.
—— —— Valerio, D., and Rowland, F.H. (1977). Comparison of Ringer lactate, dextran 70, hydroxyethyl starch, hemaccel and a colloid–crystalloid mixture in shock resuscitation. *Scot. Med. J.* **22**, 190.
Solanke, T.F. (1968*a*). Clinical trial of 6 per cent hydroxyethyl starch (a new plasma expander) *Br. med. J.* **iii**, 783.
—— (1968*b*). Clinical trial of six per cent HES in 19 patients. *W. Afr. Med. J.* **17**, 242.
—— Khwaja, M.S., and Madojemu, E.I. (1971). Plasma volume studies with four different plasma volume expanders. *J. Surg. Res.* **11**, 140.
Spencer, H.H., Starkweather, W.H., and Knorpp, C.T. (1969). Preservation of platelets with hydroxyethyl starch. *Cryobiology* **6**, 285.
Spurlin, H.M. (1939). Arrangement of substituents in cellulose derivatives. *J. Am. chem. Soc.* **61**, 222.
Srivastava, H.C. and Ramalingam, K.V. (1967). Distribution of hydroxyethyl starch. I. *Die Staerke* **19**, 295.
—— —— and Doshi, N.M. (1969). Distribution of hydroxyethyl groups in commercial hydroxyethyl starch. II. *Die Staerke* **21**, 181.
Starkweather, W.H., Knorpp, C.T., and Weatherbee, L. (1971). Changes in biochemical parameters of erythrocytes after freezing and thawing with hydroxyethyl starch as a cryoprotective agent. *Cryobiology* **8**, 392.
—— —— Spencer, H.H., and Gikas, P. (1968). The preservation of erythrocyte enzyme activity during freezing and thawing. *Cryobiology* **4**, 256.
Stiff, P.J., Clarkson, B., Zaroulis, C., and Murgo, A. (1980). Successful cryopreservation of human bone marrow using dimethylsulfoxide (DMSO) and hydroxyethyl starch (HES) as cryoprotectants. Presented to *The 18th Congress of the International Society of Haematology*. Montreal, Canada, 16–22 August.
Strauss, R.G. (1979). In-vitro comparison of the erythrocyte sedimenting properties of dextran, hydroxyethyl starch and a new low-molecular-weight hydroxyethyl starch. *Vox Sang.* **37**, 268.
—— (1981). Review of the effects of hydroxyethyl starch on the blood coagulation system. *Transfusion* **21**, 299.
—— and Koepke, J.A. (1979). Chemistry, pharmacology and donor effects of hydroxyethyl starch as used during leukapheresis. *Proc. Haem. Res. Inst., Boston, Mass.*
—— —— (1980). Chemistry, pharmacology and donor effects of hydroxyethyl starch as used during leukapheresis. *Plasma Ther.* **1**, 35.
—— —— and Maguire, L.C. (1977). Properties of neutrophils (PMN) prepared for transfusion by the Haemonetics Cell-Separator. *Blood* **50** Suppl. 1, 310.
—— —— —— and Thompson, J.S. (1979). Properties of neutrophils collected by discontinuous-flow centrifugation leukapheresis employing hydroxyethyl starch. *Transfusion* **19**, 12.
—— —— —— —— (1980*a*). A review of the clinical and laboratory effects on

donors of intermittent-flow centrifugation platelet-leukapheresis performed with hydroxyethyl starch and citrate. *Clin. Lab. Haematol.* **2**, 1.
—— —— —— —— (1980*b*). Effect of intermittent-flow centrifugation leukapheresis on donor leukocyte counts. *Acta haematol.* **63**, 128.
—— Spitzer, R.E., Stitzel, A.E., Urmson, J.R., and Maguire, L.C. (1978). Complement changes during intermittent-flow centrifugation leukapheresis (IFCL). *Blood* **52**, Suppl. 1, 304.
—— —— —— —— Spitzer, R.E., Stitzel, A.E., and Urmson, J.R. (1980*c*). Complement changes during leukapheresis. *Transfusion* **20**, 32.
Sudo, I. (1972). Clinical trial of Hespander as a new plasma expander. *Shinryo-To-Shinyaku* **9**, 1661.
—— Fukuda, Y., and Fujita, T. (1972). The effects of Hespander on body fluid distribution of splenectomized dogs. *Jap. J. Anesthiol.* **21**, 368.
Sumida, S. *et al.* (1971). Experience on hydroxyethyl starch in saline (McGaw). *Jap. J. Clin. exp. Med.* **48**, 2633.
Sunder-Plasmann, L., Reichart, B., Hugel, W., Beyer, J., and Brunner, L. (1976). Effects of dextran and hydroxyethyl starch on pulmonary circulation. *Eur. Surg. Res.* **8**, 122.
Sussman, L.N., Colli, W., Pichetshote, C. (1975). Harvesting of granulocytes using a hydroxyethyl starch solution. *Transfusion* **15**, 461.
Suyama, T. (1972). Clinical examination of Hespander (mainly in gynecological field). *Shinryo-To-Shinyaku* **9**, 2425.
—— *et al.* (1971). Transfusion effect of 6% hydroxyethyl starch in dog induced hemorrhagic shock. *Med. Consul. New Rem.* **5**, 763.
Suzuki, H. *et al.* (1971). Study on hydroxyethyl starch (HES) – Absorption, distribution and excretion of ^{14}C-HES. *Shinyaku Rinsho* **5**, 1202.
Suzuki, N. (1974). Experience of the use of Hespander in surgical patients (during operation for gastric ulcer or gallstone). *Geka Shinryo* **16**, 1269.
Tagashira, I. (1976). Clinical studies on Hespander in the aged patient. *Med. Consul. New Rem.* **13**, 2223.
Tahan, M. and Zilkha, A. (1969*a*). *J. Polymer Sci.* **7**, 1815.
—— —— (1969*b*). *J. Polymer Sci.* **7**, 1825.
Tai, H., Powers, R.M., and Protzman, P. (1966). Determination of hydroxyethyl group in hydroxyethyl starch by pyrolysis-gas chromatography technique. *Analyt Chem.* **36**, 108.
Takai, K. (1972). Studies on the in-vivo distribution and excretion of ^{14}C-labelled hydroxyethyl starch-blood residual rate, excretion and body-distribution of HES. *Report of Radioisotope Investigational Group at Kumamoto University* **7**, 91.
Takaori, M. (1966). Changes of pH of blood diluted with plasma and plasma substitutes in-vitro. *Transfusion* **6**, 597.
—— and Safar, S. (1965). Adaptation to acute hemodilution with four blood volume expanders. In *Proceedings of the Third Conference on Artificial Colloidal Agents*, p. 170. NAS-NRC, Washington, DC.
—— —— (1966). Body fluid compartment changes after hemorrhage treated with low molecular weight dextran (LMWD) and hydroxyethyl starch (HES). *Fedn Proc. Fedn Am. Socs. exp. Biol.* **25**, 698.
—— —— (1967). Treatment of massive hemorrhage with colloid and crystalloid solutions. *J. Am. med. Ass.* **199**, 297.
—— —— (1976). Critical point in progressive hemodilution with hydroxyethyl starch. *Kawasaki Med. J.* **2**, 211.
—— —— and Galla, S.J. (1968). Comparison of hydroxyethyl starch with plasma and dextrans in severe haemodilution. *Can. Anaesth. Soc. J.* **15**, 347.

—— —— —— (1970). Changes in body fluid compartments during hemodilution with hydroxyethyl starch and dextran 40. *Archs Surg.* **100**, 263.
—— —— Harris, L.C., and Loehning, R. (1965). Acute hemodilution with plasma expanders. *Anesthesiology* **26**, 261.
—— *et al.* (1971). Hemorrhage–transfusion balance during surgical operation, an observation on hematocrit alteration. *Med. Consult. New Rem.* **8**, 2391.
Takasugi, N. (1976). Clinical experience of Hespander (a new plasma expander) in patients with terminal cancer of the digestive system and the subsequent malignant fluid. *Yakubutsu Ryoho* **9**, 1281.
Takeyoshi, S. *et al.* (1971*a*). Experience on clinical application of hydroxyethyl starch (HES). *Shinyaku Rinsho* **5**, 1256.
—— Yamauchi, N., Hiyoshi, K., Ikeda, Y., and Sato, T. (1971*b*). Influences of hydroxyethyl starch solution upon circulatory dynamics, blood coagulation and erythrocyte aggregation. I. Studies on volunteers. *Jap. J. Anesthesiol.* **20**, 648.
Takiguchi, M. (1977). Effect of hydroxyethyl starch (Hespander) on blood coagulation and fibrinolysis. *Rinsho-To-Kenkyu* **54**, 2716.
Tamada, T., Okada, T., Ishida, R., and Irikura, T. (1970). Studies on hydroxyethyl starch as a plasma volume expander. I. Persistence and enzyme resistence of HES. *Pharmacometrics* **4**, 505.
—— —— —— —— and Kamishita, K. (1971). Studies on hydroxyethyl starch as a plasma expander. II. Influence of molecular weight of hydroxyethyl starch on its physicochemical and biological properties. *Chem. Pharmacol.* **19**, 286.
Tanaka, R. (1974). Clinical comparative examination of plasma substitutes (comparison of solita GL, saviosol and Hespander). *Kitasato Igaku* **4**, 382.
Tanford, C. (1961). *Physical chemistry of macromolecules*. Wiley, New York.
Thewlis, B.N. (1975). Studies on hydroxyethyl starch. III. Preparation of 2-hydroxyethyl ethers of glucose. *Die Staerke* **27**, 336.
Thompson, W.L. (1963). *Hydroxyethyl starch: a prospective plasma substitute.* University Microfilms, Ann Arbor.
—— (1965). Interaction of hydroxyethyl starch and dextran with plasma proteins and erythrocyte envelopes. In *Proceedings of the Third Conference on Artificial Colloidal Agents*, p. 36. NAS-NRC, Washington, DC.
—— (1966). Interaction of hydroxyethyl starch and dextran with plasma proteins and erythrocyte envelopes. *Biorheology* **3**, 49.
—— (1974). Plasma proteins and substitutes in critically ill patients. Scientific Exhibit, American College of Surgeons, Miami, Florida.
—— (1977*a*). Leserzuschrift. *Infusionsther. Klin. Ernaehr.* **3**, 102.
—— (1977*b*). Leserzuschrift. *Infusionsther. Klin. Ernaehr.* **3**, 301.
—— (1978). Hydroxyethyl starch. *Prog. Clin. Biol. Res.* **19**, 283.
—— (1979). Leserzuschrift. *Infusionsther. Klin. Ernaehr.* **6**, 453.
—— and Walton, R.P. (1962). Parenteral administration of hydroxyethyl starches. Presented at *The First Conference on Artififical Colloids for Intravenous Use.* NAS-NRC Conference, Washington, DC.
—— —— (1963). Blood changes, renal function and tissue storage following massive infusion of hydroxyethyl starches. *Fedn Proc. Fedn. Am. Socs exp. Biol.* **22**, 640.
—— —— (1964*a*). Elevation of plasma histamine levels in the dog following administration of muscle relaxants, opiates and macromolecular polymers. *J. Pharmacol. exp. Ther.* **143**, 131.
—— —— (1964*b*). Circulatory responses to intravenous infusions of hydroxyethyl starch solution. *J. Pharmacol. exp. Ther.* **146**, 359.

–– and Gadsden, R.H. (1965). Prolonged bleeding times and hypofibrinogenemia in dogs after infusion of hydroxyethyl starch and dextran. *Transfusion* **5**, 440.
–– Walton, R.P., and Britton, J.J. (1960). Blood levels of glucose and total carbohydrate following intravenous infusion of dextran and hydroxyethylated starches. *Fedn Proc. Fedn. Am. Socs exp Biol.* **19**, 103.
–– –– –– (1962). Persistence of starch derivatives and dextran when infused after hemorrhage. *J. Pharmacol. Exp. Ther.* **136**, 125.
–– –– and Lee-Benner, L. (1964*a*). Intravascular persistence and urinary excretion of hydroxyethyl starch solutions in normovolemic dogs. *Fedn Proc. Fedn Am. Socs exp Biol.* **23**, 539.
–– –– and Wayt, D.H. (1964*b*). Bleeding volume indices of hydroxyethyl starch, dextran, blood, and glucose. *Proc. Soc. exp. Biol. Med.* **115**, 474.
–– Bloxham, D.D., and Rudnick, M.S. (1977*a*). New short-persistence hydroxyethyl starch (HES-S): Kinetics and efficacy in dogs and patients. *Intens. Care Med.* **3**, 206.
–– –– –– (1977*b*). Short persistence hydroxyethyl starch – kinetics in dogs. *Clin. Res.* **25**, 277A.
–– Fukushima, T., Rutherford, R.B., and Walton, R.P. (1970). Intravascular persistence, tissue storage and excretion of hydroxyethyl starch. *Surg. Gynecol. Obstet.* **131**, 965.
–– –– –– –– (1979). Intravasale Persistenz, Gewebsspeicherung und Ausscheidung von Hydroxyaethyl staerke (HAES). *Infusionstherap. Klin. Ernaehr.* **6**, 243.
Thorsen, G. and Hint, H. (1954). Aggregation, sedimentation, and intravascular sludging of erythrocytes. Interrelation between suspension stability and colloids in suspension fluid. *Acta chir. scand. (Suppl.)* **154.**
Toyama, T. (1972). Clinical studies of Hespander. *Shinryo-To-Shinyaku* **9**, 1972.
Tsushima, M. *et al.* (1971). Effect of hydroxyethyl starch (HES) solution on blood volume and hematocrit. *Jap. J. clin. Exp. Med.* **48**, 2961.
Uemura, R. (1973). Basic studies of hydroxyethyl starch (Hespander) – effect on shock and function of blood coagulation and plasma expanding effect. *Kekkan-To-Myakkan* **4**, 847.
Ungerleider, R.S., Appelbaum, F.R., Breillatt, J.P., Willis, D.D., and Deisseroth, A.B. (1977). Increased yield of granulocytes using a new continuous flow centrifuge for collection. *Blood* **50** (Suppl. 1), 311.
Ulmi, W. (1979). *Volumeneffekt und Albuminkinetik von 6%iger Hydroxyaethylstaerke (HAES) bei jungen normovolamen Patienten.* Schudel, Riehen.
Ura, S. (1974). Studies on transfusion of Hespander in laparotomical patients with special regard to acid–base equilibrium, and osmotic pressure of plasma and urine. *Med. Consul. New Rem.* **11**, 1907.
Vallejos, C.S., McCredie, K.B., and Freireich, E.J. (1972). Improved leukocyte collections with hydroxyethyl starch (HES) from patients with chronic myelocytic leukemia (CML). *Clin. Res.* **20**, 503.
–– –– –– and Brittin, G.M. (1973). Biological effects of repeated leukapheresis of patients with chronic myelogenous leukemia. *Blood* **42**, 925.
–– –– –– Bodey, G.P., and Hester, J.P. (1975). White blood cell transfusion for control of infections in neutropenic patients. *Transfusion* **15**, 28.
Van der Bij, J.R. (1967). Estimation of ether groups in starch derivatives. *Die Staerke* **19**, 256.
Vinazzer, H. and Bergmann, H. (1975). Zur Beeinflussung postoperativer Aenderungen der Blutgerinnung durch Hydroxyaethylstaerke. *Anaesthesist* **24**, 517.
Vineyard, G.C., Bradley, B.E., Defalco, A., Lawson, D., Wagner, T.A., Pastis,

W.K., Nardella, F.A., and Hayes, J.R. (1966). Effect of hydroxyethyl starch on plasma volume and hematocrit following hemorrhagic shock in dogs. Comparison with dextran, plasma and Ringer's. *Ann. Surg.* **164**, 891.

von Matthiessen, H., Mempel, G., and Kolb, E. (1977/78). Anaphylaktoide Reaktion nach Hydroxyaethylstaerke. *Anaesth. Praxis* **14**, 61.

Waldman, A.A., Miller, R.S., Familletti, P., Rubenstein, S., and Pestka, S. (1982). Induction and production of interferon with human leukocytes from normal donors with the use of Newcastle disease virus. In *Methods in enzymology*. In press.

Wales, M., Marshall, P.A., and Weisberg, S.G. (1953). Intrinsic viscosity molecular weight relations for dextran. *J. Polymer Sci.* **10**, 229.

Walford, R.L. (1969). The isoantigenic systems of human leukocytes. Medical and biological significance. *Sem. Hematol.* **2**, 2.

Wallenius, G. (1953). Some procedures for dextran estimation in various body fluids. *Acta soc. med.* **59**, 69.

Walton, R.P., Richardson, J.A., and Thompson, W.L. (1959). Hypotension and histamine release following intravenous injection of plasma substitutes. *J. Pharmacol. exp. Ther.* **127**, 39.

—— Hauck, A.L., and Herman, E.H. (1966). Structural specificity of dextran in producing anaphylactoid reactions in rats. *Proc. Soc. exp. Biol. Med.* **121**, 272.

Watzek, C., Wagner, O., Draxler, V., Gilly, H., Schwarz, S., Sporn, P., Steinbereithner, K., and Zekert, F. (1978). Influence of normovolaemic hemodilution with hydroxyethyl starch on circulation and organ function in vascular surgery. *Wien Klin. Wochenschr.* **90**, 224.

Weatherbee, L., Allen, E.D., and Permoad, P.A. (1979). Full units of RBCs frozen with HES 150/0.70. *Vox Sang.* **37**, 362.

—— Spencer, H.H., Knorpp, C.T., and Lindenauer, S.M. (1972). Method for rapid freezing and thawing of full unit quantities of packed red blood cells with hydroxyethyl starch in plastic bags. *Cryobiology* **9**, 317.

—— —— Allen, E.D., Lindenauer, S.M., and Permoad, P.A. (1974*a*). Review of hydroxyethyl starch as an extracellular cryoprotective agent of red bloodcells. *Cryobiology* **11**, 537.

—— —— —— —— —— (1974*b*). Effect of plasma on red blood cells frozen with hydroxyethyl starch (HES). *Cryobiology* **11**, 538.

—— —— —— —— —— (1975*a*). Effect of plasma on hydroxyethyl starch-preserved red cells. *Cryobiology* **12**, 119.

—— —— —— —— —— (1975*b*). Red-cells preserved with 10 per cent hydroxyethyl starch – effect of prefreezing washing. *Cryobiology* **12**, 513.

—— —— Knorpp, C.T., Lindenauer, S.M., Gikas, P., and Thompson, N.W. (1974*c*). Coagulation studies after transfusion of hydroxyethyl starch protected frozen blood in primates. *Transfusion* **14**, 109.

Wheeler, T.G., McCredie, K.B., Freireich, E.J., and Daniels, T.V. (1974). Increased efficiency of leukocyte collection from patients with chronic myelocytic leukemia. *Transfusion* **14**, 253.

Wiedersheim, M. (1957). An investigation of oxyethyl-starch as a new plasma volume expander in animals. *Archs Int. Pharmacodyn.* **111**, 353.

Winton, E.F. and Vogler, W.R. (1978). Development of a practical oral dexamethasone premedication scheduling leading to improved granulocyte yields with the continuous-flow centrifugal blood cell separator. *Blood* **52**, 249.

Wong, S.C. and Rock, G. (1979). The function of platelets in hydroxyethyl starch: Efficiency of combined platelet–leukocyte preparations. *Blood* **54** Suppl. 1, 130a.

Woods, A.H., Gibbs, R., and Holmberg, A. (1975). The anaemia associated with repeated leucapheresis. In *Leucocytes: separation, collection and transfusion* (ed. J.M. Goldman and R.M. Lowenthal). Academic Press, London.
Woods, K.R., Horowitz, B., Wiebe, M., and Waldman, A.A. (1980). Quarterly Program Report No. 1, National Heart, Lung, and Blood Institute, NIH.
Wusteman, M.C. (1978). Comparison of colloids for use in isolated normothermic perfusion of rabbit kidneys. *J. Surg. Res.* **25**, 54.
Yamada, T. and Sakamoto, T. (1975). Hemodynamic changes following rapid infusion of hydroxyethyl starch. *Jap. J. Anesthesiol.* **24**, 891.
Yamamoto, E. *et al.* (1974). Clinical study of hydroxyethyl starch (6-HES). *Jap. J. Anesthesiol.* **24**, 570.
Yamamoto, K. (1972). Clinical experience with Hespander (in surgical patients in obstetric, orthopedic, neurosurgical and gastrointestinal fields). *Yakubutsu Ryoho* **5**, 1313.
Yamamoto, T. and Momose, T. (1972). Clinical evaluation of hydroxyethyl starch. *Iryo* **26**, 118.
Yamasaki, H. (1973). The colloid osmotic pressure of normal human blood, dextran and hydroxyethyl starch. *Jap. J. Anesthesiol.* **22**, 1349.
—— (1974). Hyper-, normo-, and hypocolloid osmotic pressure effects of colloid solutions on human subjects. *Jap. J. Anesthesiol.* **23**, 812.
—— (1975). A comparative study on renal effects of dextran 40 and HES 40 administered to dehydrated rabbits. *Jap. J. Anesthesiol.* **24**, 580.
Yamashina, H. (1975). Clinical experience of hydroxyethyl starch solution (Hespander) in the acute phase of cardiac infarction. *Med. Consul. New Rem.* **12**, 1905.
Yara, I. (1974). Clinical studies on hemodilution during extracorporeal circulation. Examination of hydroxyethyl starch as a plasma substitute. *Geka Shinryo* **16**, 591.
Yoshida, M., Yamashita, T., Matsuo, J., and Kishikawa, T. (1973). Enzymatic degradation of hydroxyethyl starch. I. Influence of the distribution of hydroxyethyl groups on the enzymic degradation of hydroxyethyl starch. *Die Staerke* **25**, 373.
Yoshikawa, H. *et al.* (1974). Effect of normovolaemic anemia with HES on distribution of cardiac output in dogs. *Jap. J. Anesthesiol.* **24**, 12.
Yoshitake, T. *et al.* (1971). Clinical trial of hydroxyethyl starch (HES) in open heart surgery. *J. New Rem. Clin.* **20**, 1413.
Zaffiri, O., Alessio-Verni, A., Pecchiari, V., and Mastroianni, A. (1969). A new plasma expander: Hydroxyethyl starch. *Minerva Anesthesiol.* **35**, 108.
Zaroulis, C.G., Leiderman, I.Z., and Lee, S.C. (1978). Successful freeze-preservation of human granulocytes. *Blood* **52** (Suppl. 1), 305.
—— —— (1980). Successful freeze-preservation of human granulocytes. *Cryobiology* **17**, 311.
Ziese, W. (1943). Beitrag zur Spezifitaet der Amylasen. I. Einwirkung von Amylasen auf Oxyaethylstaerke. *Hoppe Seylers Z. Physiol. Chem.* **229**, 213.
—— (1935). Beitrag zur Spezifitaert der Amylasen. II. Über das Nichtauftreten von Redukfoinsvermoegen beim enzymatischen Abbau von aus Nativstaerke gewonnener Oxyaethylstaerke. *Hoppe Seylers Z. Physiol. Chem.* **235**, 235.
Zimmermann, W.E. (1971). Hydroxyethyl starch compared with other plasma substitutes in the treatment of hemorrhagic shock. In *Les solutes de substitution.* Librairie Arnette, Paris.
—— and Bannert, P. (1970). Studies on the action of hydroxyethyl starch. In *Shock, metabolic disorders and therapy* (ed. W.E. Zimmerman and I. Staub). Schattauer, Stuttgart.

—— and Rehfeld, K.H. (1971). Function and importance of the RES-system during perfusion with different solutions for the preservation and transplantation of the liver. *J. Reticuloendoth. Soc.* **23**, 197.

Index